普通高等教育"十三五"规划教材

计算机应用基础（双语版）

主　编　于鲁佳

副主编　张　宇　董燕妮

内 容 提 要

本书根据教育部制定的《高等学校非计算机专业计算机课程基本要求》并结合目前对外教育及高校学生双语教学的内容要求为依据编写。

全书主要介绍计算机基本知识和理论，实际操作和应用。内容包括计算机的基本概念和发展情况，Windows 操作系统基础，常见的 Office 软件 Word、Excel 和 PowerPoint 的应用方法，以及网络基础知识和网络安全及网络最新技术的简介。

本书注重理论和实际相结合，实用性和操作性强。本书适合作为各类高等学校计算机教学的双语教学使用，也可作为国际交流学生计算机基础自学参考书。

本书配有电子教案，读者可以从中国水利水电出版社网站和万水书苑免费下载，网址为：http://www.waterpub.com.cn/softdown 和 http://www.wsbookshow.com。

图书在版编目（CIP）数据

计算机应用基础 : 双语版 / 于鲁佳主编. -- 北京 : 中国水利水电出版社，2016.9（2024.9 重印）
普通高等教育"十三五"规划教材
ISBN 978-7-5170-4671-4

Ⅰ. ①计… Ⅱ. ①于… Ⅲ. ①电子计算机－高等学校－教材 Ⅳ. ①TP3

中国版本图书馆CIP数据核字(2016)第207821号

策划编辑：石永峰　　责任编辑：邓建梅　　封面设计：李　佳

书　名	普通高等教育"十三五"规划教材 计算机应用基础（双语版） JISUANJI YINGYONG JICHU (SHUANGYU BAN)
作　者	主　编　于鲁佳 副主编　张　宇　董燕妮
出版发行	中国水利水电出版社 （北京市海淀区玉渊潭南路 1 号 D 座　100038） 网址：www.waterpub.com.cn E-mail：mchannel@263.net（答疑） 　　　　sales@mwr.gov.cn 电话：（010）68545888（营销中心）、82562819（组稿）
经　售	北京科水图书销售有限公司 电话：（010）68545874、63202643 全国各地新华书店和相关出版物销售网点
排　版	北京万水电子信息有限公司
印　刷	三河市鑫金马印装有限公司
规　格	184mm×260mm　16 开本　12.75 印张　312 千字
版　次	2016 年 9 月第 1 版　2024 年 9 月第 4 次印刷
印　数	5001—6000 册
定　价	45.00 元

凡购买我社图书，如有缺页、倒页、脱页的，本社营销中心负责调换

版权所有·侵权必究

前　　言

随着计算机技术和网络科技的不断发展，计算机已经成为人们获取知识、传播知识的重要载体。且随着国际交流的不断深入，越来越多的学生更加注重"全球化"的教育。本教材是针对大学学习计算机基础知识的双语教材。本教材采用通俗易懂的方式介绍了计算机的基本知识，常用软件的使用以及计算机网络的基本使用方法和问题。

本书共分为六章。第一章介绍了计算机的基本知识和概念，包含了计算机的发展历史、基本组成，进制转换等基本方法和概念。第二章介绍了 Windows 10 操作系统，由于 Microsoft 公司现在主推 Windows 10 操作系统，本书选择该系统作为主要介绍，并引入其他前期操作系统，并介绍他们的异同点。主要介绍 Windows 10 操作系统的桌面管理，任务栏使用，窗口管理，文件及文件夹管理，系统管理，多种附件的使用，以及最新的 Cortana 和 Math Input Panel 等功能。第三章主要介绍了 Office 2010 文件包中的 Word，重点介绍了 Word 的编辑和排版，表格的制作和编辑，图片的插入及编辑等知识和技巧。第四章主要介绍了电子表格 Excel 的使用，重点介绍了表格的基本功能，表格中公式及函数的使用方法，对数据的处理，并引入了大量的实例来解决地址等易混淆的问题。第五章介绍了幻灯片 PowerPoint 的制作方法，从简单的幻灯片的新建、制作管理，到动画的应用，以便于初学者根据需要进行由浅入深的学习。第六章介绍了现阶段常见的网络知识，如网络的基本概念和结构、网络的通用设备，重点介绍了网络安全的问题和解决方法。帮助读者有针对性的了解现阶段网络的优势和问题。整本书的着重点是帮助学生在学习计算机基础知识的同时，提升自我计算机专业英语应用的能力。

本书的第 1、2 章由张宇编写；第 3、4、6 章由于鲁佳编写；第 5 章由董燕妮编写。

由于时间仓促及作者水平有限，书中难免有一定的疏漏和问题，恳请广大读者批评指正。

编　者

2016 年 6 月

目 录

前言

Chapter 1　Fundamental of Computer
　　计算机历史 ·· 1
　1.1　An Introduction of computer ············· 1
　　1.1.1　Computer's History ···················· 1
　　1.1.2　An Introduction of computer ······ 1
　　1.1.3　Computers' classified ················· 3
　1.2　Numbering System ······························· 9
　　1.2.1　An overview of the Numbering System · 9
　　1.2.2　Binary Code ······························· 9
　　1.2.3　Other Numbering System ··········· 9
　　1.2.4　Conversion between Different Numbering Systems ··············· 10
　1.3　Computer System ······························· 12
　　1.3.1　Hardware ···································· 12
　　1.3.2　Software ····································· 18
　1.4　Reference ··· 20
　1.5　English-Chinese Key Terms ················ 20

Chapter 2　Operating System
　　操作系统 ·· 21
　2.1　An Introduction of Operating System ········ 21
　　2.1.1　What is an Operating System ········· 21
　　2.1.2　Operating System Classification ····· 21
　　2.1.3　Management of Files ··················· 27
　　2.1.4　Work with Control Panel ············ 34
　2.2　Accessories ··· 40

Chapter 3　Microsoft Word 2010
　　文字处理软件——Word 2010 ················ 45
　3.1　An introduction of MS Word 2010 ····· 45
　　3.1.1　Word 2010 Components ············· 45
　　3.1.2　Word 2010 Layouts ···················· 47
　3.2　Creating a document ·························· 48
　　3.2.1　Creating a new document ·········· 49
　　3.2.2　Using Template ·························· 51

　　3.2.3　Saving a document ···················· 52
　　3.2.4　Closing a document ··················· 53
　3.3　Formatting a document ······················ 54
　　3.3.1　Editing the text ·························· 55
　　3.3.2　Document Setting ······················ 60
　　3.3.3　Pictures and Text ······················· 68
　　3.3.4　Table ··· 72
　3.4　Finalizing a Document ······················· 79
　　3.4.1　Page Design ······························· 79
　　3.4.2　Preview and Print ······················ 84
　　3.4.3　PDF Conversion ························· 84

Chapter 4　Microsoft Excel 2010
　　电子表格 Excel 2010 ······························ 86
　4.1　An introduction of MS Excel 2010 ······ 86
　　4.1.1　Excel 2010 Components ············· 86
　　4.1.2　Start an Excel ···························· 88
　4.2　Creating a Worksheet ························· 92
　　4.2.1　Creating a new Worksheet ········· 92
　　4.2.2　Fill in Data ································· 94
　　4.2.3　Closing a Worksheet ················ 100
　4.3　Formulas and Functions ··················· 100
　　4.3.1　Formulas ·································· 102
　　4.3.2　Functions ································· 104
　4.4　Formatting a Worksheet ··················· 111
　　4.4.1　Editing the cells ······················· 112
　　4.4.2　Borders and Shadings ·············· 115
　4.5　Finalizing a Worksheet ····················· 117
　　4.5.1　Sorting data ····························· 117
　　4.5.2　Filtering data ··························· 119
　　4.5.3　Headers and Footers ················ 120
　　4.5.4　Preview and Print ···················· 122
　4.6　Creating a chart ································ 123
　　4.6.1　Creating a new chart ················ 124

4.6.2	Modifying a chart	125
4.6.3	Formatting and editing a chart	125

Chapter 5　Microsoft PowerPoint 2010
　　　　　　电子幻灯片的制作——
　　　　　　PowerPoint 2010 ················ 127

- 5.1 An overview of PowerPoint 2010 ········ 127
  - 5.1.1 PowerPoint 2010 Components ········ 127
  - 5.1.2 Slides views ·············· 128
- 5.2 Creating and Formatting Slides ············ 136
  - 5.2.1 Creating new Slide ·············· 136
  - 5.2.2 Contents input and edit ············ 138
  - 5.2.3 Choosing Layout ··············· 140
  - 5.2.4 Choosing Themes ·············· 141
- 5.3 Formatting Slides ·············· 142
  - 5.3.1 Text ····················· 142
  - 5.3.2 Pictures ·················· 144
  - 5.3.3 Table ····················· 146
- 5.4 Showing Effects ················ 150
  - 5.4.1 Animations ················ 150
  - 5.4.2 Hyperlink ················· 152
  - 5.4.3 Music ···················· 153
- 5.5 Presenting Slides ··············· 155
  - 5.5.1 Slide show ················· 155
  - 5.5.2 Context Help ················ 156

Chapter 6　Online Connection
　　　　　　网络技术 ·················· 157

- 6.1 Brief introduction of Network ············ 157
  - 6.1.1 Network Concept ············· 157
  - 6.1.2 History of Network ············ 158
  - 6.1.3 Network Classification ········· 160
  - 6.1.4 Network Protocol ············· 165
- 6.2 Local Area Network (LAN) ············· 167
  - 6.2.1 Structure of LAN ············· 168
  - 6.2.2 LAN Components ············· 172
- 6.3 Internet acknowledges ············· 174
  - 6.3.1 Introduction of the Internet ········ 174
  - 6.3.2 Client/Server format ············ 175
  - 6.3.3 TCP/IP Protocol ·············· 176
  - 6.3.4 IP address ················· 176
- 6.4 Internet Application ················ 179
  - 6.4.1 World Wide Web ············· 179
  - 6.4.2 E-mail ···················· 180
  - 6.4.3 File download ··············· 185
  - 6.4.4 Search on Internet ············· 186
- 6.5 Network Security ················ 188
  - 6.5.1 Security Attack ··············· 189
  - 6.5.2 Internet Virus ················ 190
  - 6.5.3 Anti-Virus Software ············· 191
- 6.6 New in Internet ················· 192
  - 6.6.1 Cloud ···················· 192
  - 6.6.2 Big Data ··················· 194

Chapter 1　Fundamental of Computer
计算机历史

1.1　An Introduction of computer

A **computer** is a general-purpose device that can be programmed to carry out a set of arithmetic or logical operations automatically. Since a sequence of operations can be readily changed, the computer can solve more than one kind of problem.[1]

According to modern societies' requirements, Internet becomes one of the most important current technologies. And computers is the most important carrier. With the advent of the Internet and higher bandwidth data transmission, programs and data that are part of the same overall project can be distributed over a network and embody the Sun Microsystems slogan: "The network is the computer."

1.1.1　Computer's History

Most histories of the modern computer begin with the Analytical Engine envisioned by Charles Babbage following the mathematical ideas of George Boole, the mathematician who first stated the principles of logic inherent in today's digital computer. Babbage's assistant and collaborator, Ada Lovelace, is said to have introduced the ideas of program loops and subroutines and is sometimes considered the first programmer. Apart from mechanical calculators, the first really usable computers began with the vacuum tube, accelerated with the invention of the transistor, which then became embedded in large numbers in integrated circuits, ultimately making possible the relatively low-cost personal computer.

Modern computers inherently follow the ideas of the stored program laid out by John von Neumann in 1945. Essentially, the program is read by the computer one instruction at a time, an operation is performed, and the computer then reads in the next instruction, and so on.

1.1.2　An Introduction of computer

The first computer in the world was born in 15th, February, 1946. The computer was named as ENIAC, is shown in Fig 1.1 (Electronic Numerical Integrator and Computer), at the U.S. Army's Aberdeen Proving Ground in Maryland.

Whole machine was built in a metal cabinet, which weighed 30 tons and was eight feet high, three feet deep, and 100 feet long. The Computers were using vacuum tube technology, the system contained over 18,000 vacuum tubes that were cooling by 80 air blowers, Fig 1.2.

Fig 1.1   the ever-first computer ENIAC

Fig 1.2   ENIAC's vacuum tubes

Programs were loaded into memory manually using switches, punched cards, or paper tapes (see Fig1.3 and 1.4).

Fig 1.3   ENIAC: coding by cable connections

Chapter 1 Fundamental of Computer
计算机历史

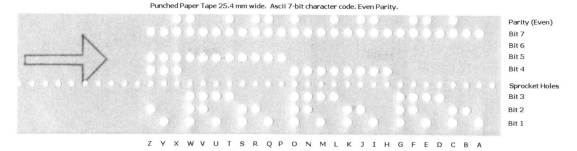

Fig 1.4　Punch card

1.1.3　Computers' classified

After ENIAC appeared, computers were developing faster and faster. Base on the main contents and inner components, computers are classified as five generations.

- First Generation Computers

The first generation computers employed during the period 1940—1956, which used the vacuum tubes technology for calculation as well as for storage and control purpose. Using vacuum tubes brought some advantages in that period:

(1) Fastest computing devices of their time, for example, ENIAC can control addition 5,000 times/sec.

(2) These computers were able to execute complex mathematical problems in an efficient manner.

Also, the disadvantages had to be considered:

(1) The functioning of these computers depended on the machine language.

(2) There were generally designed as special-purpose computers.

(3) The use of vacuum tube technology makes these computers very large and bulky.

(4) They were not easily transferable from one place to another due to their huge size and also required to be placed in cool places.

(5) They were single tasking because they could execute only one program at a time.

(6) The generated huge amount of heat and hence were prone to hardware faults.

- Second Generation Computers

The second generation computers employed during the period 1956—1963, which used transistors in place of vacuum tubes in building the basic logic circuits.

The advantages of this generation were:

(1) Fastest computing devices of their time;

(2) Easy to program because of the use assembly language;

(3) Could be transferred from one place to other very easily because they were small and light;

(4) Required very less power in carrying out their operations;

(5) More reliable, did not require maintenance at regular intervals of time.

 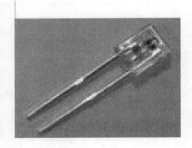

Fig 1.5   Transistors

This generation had some disadvantages:

(1) The input and output media were not improved to a considerable extent

(2) Required to be placed in air-conditioned places

(3) The cost of these computers was very high and they were beyond the reach of home users

(4) Special-purpose computers and could execute only specific applications

- Third Generation Computers

Generally, the third generation was called Integrated Circuits era and employed during the period 1964—1975.

The advantages of 3rd generation were:

(1) Fastest computing devices;

(2) Very productive;

(3) Easily transportable from one place to another because of their small size;

(4) Use high-level languages;

(5) Could be installed very easily and required less space;

(6) Can execute any type of application;

(7) More reliable and require less frequent maintenance schedules.

They also provided so many disadvantages, such as :

(1) The storage capacity of these computers was still very small;

(2) The performance of these computers degraded while executing large applications, involving complex computations because of the small storage capacity;

(3) The cost of these computers was very high;

(4) They were still required to be placed in air-conditioned places.

- Fourth Generation Computers

The fourth generation computers employed during 1975—1989 and people are known this generation as Large Scale Integration technology and Very Large Scale Integration technology. Meanwhile, the term Personal Computer (PC) became known to the people during this era.

Compare with other generation, this era's advantages were:

(1) Very powerful in terms of their processing speed and access time;

(2) Storage capacity was very large and faster;

(3) Highly reliable and required very less maintenance;
(4) User-friendly environment;
(5) Programs written on these computers were highly portable;
(6) Versatile and suitable for every type of applications;
(7) Require very less power to operate.

And the disadvantages were:

(1) The soldering of LSI and VLSI chips on the wiring board was not an easy task and required complicated technologies to bind these chips on the wiring board;

(2) The working of these computers is still dependent on the instructions given by the programmer.

- Fifth Generation Computers

The different types of modern digital computers come under this category.

Use Ultra Large Scale Integration technology that allows almost ten million electronic components to be fabricated on one small chip. Therefore, this category brings so many benefits rather than disadvantages, such as:

(1) Fastest and powerful computers till date;
(2) Being able to execute a large number of applications at the same time and that too at a very high speed;
(3) Decreasing the size of these computers to a large extent;
(4) The users of these computers find it very comfortable to use them because of the several additional multimedia features;
(5) They are versatile for communications and resource sharing.

Sometimes, people define computers' category according to the following three criteria: operating principles, applications, or size and capability.

If people only consider on operating principles, all computers can separate as 3 different types.

- **Analog computers**, which represent data in the form of continuous electrical signals having a specific magnitude.
- **Digital computers**, which generally store and process data with the digital format.
- **Hybrid computers**, which are a combination between analog computer and digital computer for presenting the best features of those 2 types of computer.

If based on applications, there are **General purpose computers** (work in all environments) and **Special purpose computers** (perform only a specified task).

Base on their abilities, they classified computers into supercomputer, mainframe computer, server, and personal computer (well known as PC, which includes handheld computer, desktop computer, notebook computer and tablet computer).

For general usage, a **personal computer** (PC) is a microcomputer designed for use by one person at a time. Prior to the PC, computers were designed for (and only affordable by) companies who attached terminals for multiple users to a single large computer whose resources were shared among all users. The term "PC" has been traditionally used to describe an "IBM-compatible" personal computer in contradistinction to an Apple Macintosh computer. Although the distinctions have become less clear-cut in recent years, people often still categorize a personal computer as either a PC or a Mac.

A **handheld computer** is shown in Fig 1.6. Usually the computer contains small keyboard or touch-sensitive screen and is designed small enough to fix into a pocket. Also, it is wireless and easy to hold when users are using. People can use handheld computer with some basic functions: appointment book, address book, calculator, and notepad.

Fig 1.6   Image for handheld computer

A **desktop computer**(see Fig 1.7) is commonly used in the public, which contains monitor, keyboards, mouse, and a main cabinet. This type of computer is popular for offices, schools, and homes.

Fig 1.7   Image for desktop computer

A **notebook computer** (or called laptop) (see Fig 1.8) is a small lightweight personal computer that incorporates a screen, a keyboard, storage devices, and processing components into a single portable unit. Notebook computer can run on power supplied by an electrical outlet or a battery.

A **tablet computer**(see Fig 1.9) is a recently popular portable computing device featuring a touch-sensitive screen. A tablet configuration lacks of a keyboard and resembles a high-tech clipboard. When tablet computers were firstly come to the public in 2002, they were significantly more popular than every other type of computers for normal user.

Fig 1.8 Image for notebook computer

Fig 1.9 Image for tablet computer

A **supercomputer**(see Fig 1.10) is a computer with a high-level computational capacity compared to a general-purpose computer. Performance of a supercomputer is measured in Floating-point Operations Per Second (FLOPS) instead of Million Instructions Per Second (MIPS). As of 2015, there are supercomputers which can perform up to quadrillions of FLOPS.[2]

Supercomputers play an important role in the field of computational science, and are used for a wide range of computationally intensive tasks in various fields, including quantum mechanics, weather forecasting, climate research, oil and gas exploration, molecular modeling (computing the structures and properties of chemical compounds, biological macromolecules, polymers, and crystals), and physical simulations (such as simulations of the early moments of the universe, airplane and spacecraft aerodynamics, the detonation of nuclear weapons, and nuclear fusion). Throughout their history, they have been essential in the field of cryptanalysis.[3]

Fig 1.10 IBM's Blue Gene Supercomputer

In current situation, super computers' quantity can be considered as the modern technology measurements. Until 2014, the super computers' numbers of each country are shown below:

A **mainframe** (also known as "big iron") is a high-performance computer used for large-scale computing purposes that require greater availability and security than a smaller-scale machine can offer. Historically, mainframes have been associated with centralized rather than distributed computing, although that distinction is blurring as smaller computers become more powerful and mainframes become more multi-purpose (see Fig 1.1 and 1.12).

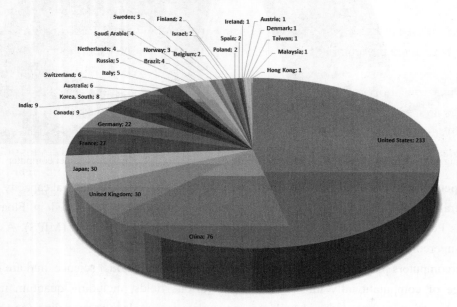

Fig 1.11　Supercomputer in different countries June, 2014

Fig 1.12　IBM Z9 Mainframe

In information technology, a **server** is a computer program that provides services to other computer programs (and their users) in the same or other computers. The purpose of a server is to "serve" data to computers connected to a network. Despite impressive performance a server can offer, these servers usually cannot use for entertainments such as sound or DVD players, and other fun accessories, so they are not suitable for a personal computers.

1.2 Numbering System

Numbering system is used by people every day. The most common one is the decimal system, which also has been known as Arabic number system.

1.2.1 An overview of the Numbering System

At present, there are so many different numbering systems have been used in several of fields. The Binary Code, Octal Decimal Number and Hexadecimal Number are offering in computer world.

Since we were little kids, we started the decimal numeral system and which is a based on the Arabic numeral system.

In the decimal numbers, it contains 10 basic numbers:0、1、2、3、4、5、6、7、8、9.

It uses positional notation and uses same symbols for different orders of magnitude, but in different places. For example, ones place, tens place, hundreds place.

All numbers can be presented as $10^0, 10^1, 10^2, 10^3$, etc. Also, the decimals can be shown as 10^{-1}, 10^{-2}. For example: In decimal, $(2317.87)_{10}$ means:

$(2317.87)_{10} = 2*10^3 + 3*10^2 + 1*10^1 + 7*10^0 + 8*10^{-1} + 7*10^{-2}$

Therefore, as a decimal number (present as D) we can follow the rule.

$(D)_{10} = (D_{n-1}D_{n-2}……D_1D_0.D_{-1}D_{-2}……D_{-m})_{10}$
$= D_{n-1}*10^{n-1} + D_{n-2}*10^{n-2} +……+ D_1*10^1 + D_0*10^0 + D_{-1}*10^{-1} + D_{-2}*10^{-2} +……+ D_{-m}*10^{-m}$

1.2.2 Binary Code

Since computer's born, they only know two numbers 0 and 1. People bring the binary numbering system to communicate with them. So in the binary system, it is a based on the binary representation (0, 1). Similar with decimal numbering system, it also uses positional notation and uses the same symbols for different orders of magnitude, but in different places, e.g., ones place, twos place, fours place.

All numbers can present as $2^0, 2^1, 2^2, 2^3$, etc. For example: $(10010.11)_2$ means

$(10010.11)_2 = 1*2^4 + 0*2^3 + 0*2^2 + 1*2^1 + 0*2^0 + 1*2^{-1} + 1*2^{-2}$
$= 16 + 0 + 0 + 2 + 0 + 0.5 + 0.25$
$= (18.75)_{10}$

So, for any binary code B, we can present as:

$(B)_2 = (B_{n-1}B_{n-2}……B_1B_0.B_{-1}B_{-2}……B_{-m})_{10}$
$= B_{n-1}*10^{n-1} + B_{n-2}*10^{n-2} +……+ B_1*10^1 + B_0*10^0 + B_{-1}*10^{-1} + B_{-2}*10^{-2} +……+ B_{-m}*10^{-m}$

1.2.3 Other Numbering System

Normally, any numbering system N can be presented as:

$(N)m = (N_{n-1}N_{n-2}……N_1N_0.N_{-1}N_{-2}……N_{-m})_m$
$= N_{n-1}*m^{n-1} + N_{n-2}*m^{n-2} +……+ N_1*m^1 + N_0*m^0 + N_{-1}*m^{-1} + N_{-2}*m^{-2} +……+ N_{-m}*m^{-m}$

In this system, N_i can be present as any numbers, such as $0.1.2\cdots\cdots m$; N and m are integer. When m=2, 8, 16, or 10, the numbering systems are Binary code, Octal numbers, Hexadecimal numbers and Decimal numbers.

1.2.4 Conversion between Different Numbering Systems

(1) Converting from other numbering systems to Decimal.

Any numbering systems are easy to convert to decimal numbers by multiply power series, for example:

$(1001.101)_2 = 1*2^3 + 0*2^2 + 0*2^1 + 1*2^0 + 1*2^{-1} + 0*2^{-2} + 1*2^{-3}$
$= 8+0+1+0.5+0+0.125$
$= 9.625$

$(56)_8 = 5*8^1 + 6*8^0$
$= 40+6$
$= 46$

$(3CF.BA)_{16} = 3*16^2 + C*16^1 + F*16^0 + B*16^{-1} + A*16^{-2}$
$= 768+12*16+15*1+11*0.0625+10*0.00390625$
$= 975+0.06640625$
$= (975.06640625)_{10}$

(2) Converting from Decimal numbering system to other numbering system.

- Decimal transfer to Binary code

Decimal number transfer to Binary code should follow these rules: calculate into integer part and decimal part. Integer part should be divided by 2 and keep reminders, and decimal parts should be multiplied by 2 and keep integers. For example:

We convert $(85.35)_{10}$ to Binary code.

First of all, we calculate the integer part $(85)_{10} = (a_{n-1} a_{n-2} \ldots a_1 a_0)_2$

Division	Quotient	Remainder	
85/2	42	1	a_0
42/2	21	0	a_1
21/2	10	1	a_2
10/2	5	0	a_3
5/2	2	1	a_4
2/1	0	1	a_5

Secondly, we calculate the decimal part $(0.135)_{10} = (a_{-1} a_{-2} \ldots a_{-m})_2$

Multiplication	Fraction	Integer	
0.35*2	0.7	0	a_{-1}
0.7*2	0.4	1	a_{-2}
0.4*2	0.8	0	a_{-3}
0.8*2	0.6	1	a_{-4}

And then, we add those two results $(85.35)_{10} = (110101.0101)_2$

- Decimal convert to Octal and Hexadecimal number

Similar with conversion between decimal and binary code, decimal convert to octal number should be divided by 8 and keep the reminders and decimal parts also should be multiplied by 8 and keep integers. For hexadecimal numbers, we use 16 for division and multiplication.

(3) Conversion between Binary to octal numbers

- Conversion between Binary to octal numbers

The converting rule is separating in two parts base on the decimal point. Before the decimal point, the integer part should count 3 digits as a group from right to left. If the left digits cannot have 3 digits, we use 0 to fit in. The decimal part also counts 3 digits as a group from left to right. If the left digits cannot have 3 digits, we use 0 to fit in. For example, we convert $(110000110101.0010001011)_2$ to an octal number.

The integer part is $(1110000110101)_2$.

```
001   110   000   110   101
 ↓     ↓     ↓     ↓     ↓
 1     6     0     6     5
```

The decimal part is $(.0010001011)_2$

```
001   000   101   100
 ↓     ↓     ↓     ↓
 1     0     5     4
```

So, we have the result $(16065.1054)_8$

- Conversion between octal to Binary numbers

We can transfer each octal digit to 3 binary digits, for example: $(1765.03)_8$.

```
 1    7    6    5 .  0    3
 ↓    ↓    ↓    ↓    ↓    ↓
001  111  110  101. 000  011
```

(4) Conversion between Binary to Hexadecimal numbers

- Conversion between Binary to Hexadecimal numbers

The converting rule is similar with octal number and separating in two parts base on the decimal point. Before the decimal point, the integer part should count 4 digits as a group from right to left. If the left digits cannot have 4 digits, we use 0 to fit in. The decimal part also counts 4 digits as a group from left to right. If the left digits cannot have 4 digits, we use 0 to fit in. For example, we use same number $(110000110101.0010001011)_2$ to transfer a hexadecimal number.

The integer part is $(1110000110101)_2$.

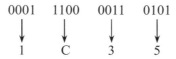

The decimal part is $(.0010001011)_2$

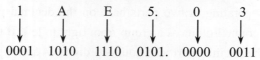

Even though we use same binary code, we receive different result $(1C35.22C)_8$

- Conversion between Hexadecimal to Binary numbers

We can transfer each hexadecimal digit to 4 binary digits, for example : $(1AE5.03)_{16}$.

```
  1      A      E      5.     0      3
  ↓      ↓      ↓      ↓      ↓      ↓
0001   1010   1110   0101.  0000   0011
```

For better calculation, we provide a table (Table 1.1).

Table 1.1

Decimal	0	1	2	3	4	5	6	7	8	9	10	11	12	13	14	15	16
Binary	0	1	10	11	100	101	110	111	1000	1001	1010	1011	1100	1101	1110	1111	10000
Octal	0	1	2	3	4	5	6	7	10	11	12	13	14	15	16	17	20
Hexadecimal	0	1	2	3	4	5	6	7	8	9	A	B	C	D	E	F	10

1.3 Computer System

A computer system is consisting of hardware and software. Also, data and human beings are indispensable parts of computer system.

1.3.1 Hardware

Computer hardware (usually simply called **hardware** when a computing context is concerned) is the collection of physical elements that constitutes a computer system. Computer hardware is the physical parts or components of a computer, such as the monitor, mouse, keyboard, computer data storage, hard disk drive (HDD), graphic cards, sound cards, memory, motherboard, and so on, all of which are physical objects that are tangible.

Hardware consists of different parts, which can be shown as below:

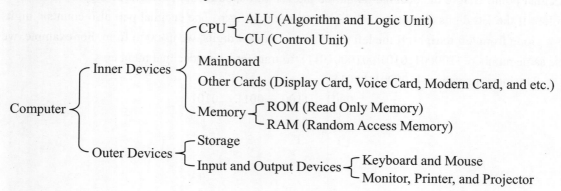

1. Processor

The processor is the brain of a computer, which is the logic circuitry that responds to and processes the Basic instructions that drives a computer. The term processor has generally replaced the term Central Processing Unit (CPU). The processor in a personal computer or embedded in small devices is often called a microprocessor (see Fig 1.13).

A processor combines 2 units, one is ALU, and the other is CU. The ALU controls all the logical process, for example AND, OR, XOR. The CU mainly charges with execution of the instructions, and controls the data sequence of in and out. The result of these routed data movements through various digital circuits (sub-units) within the processor produces the manipulated data expected by a software instruction (loaded earlier, likely from memory).

When data and instruction are waiting for ALU and CU to process, there are some small units can store them temporarily. Those small units also called as register. Depending on the processors, the registers may store 8, 16, 32, or 64 bits. Those storage sizes control processors' performance, which also call words. Therefore, we have n-bits processors, such as 32-bits processor or 64-bits processor.

Fig 1.13   a CPU processor

2. Mainboard

Mainboard (Motherboard) is shown below. Since Personal Computer (PC) has invented, mainboard is the most important part to connect each computer parts (see Fig 1.14).

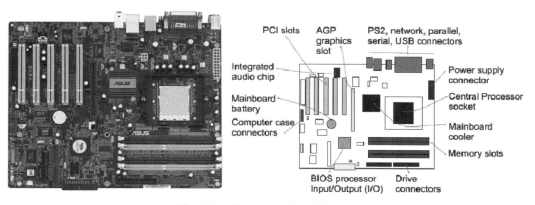

Fig 1.14   The image of mainboard

The mainboard consists of:
- Central Processor Sockets (or slots) in which one or more microprocessors may be installed. In the case of CPU's in ball grid array packages, such as the VIA C3, the CPU is directly solidified to the motherboard.[4]
- Memory slots into which the system's man memory is to be installed (typically in the form of DIMM modules containing DRAM chips).
- A clock generator which produces the system clock signal to synchronize the various components, which connect with mainboard battery and charge with system timing.
- Slots of expansion cards (for example the interface to the system via the buses supported by the chipset).
- Power connectors, which receive electrical power from the computer power supply and distribute it to the CPU, chipset, main memory, and expansion cards. As of 2007, some graphics cards (e.g. GeForce 8 and Radeon R600) require more power than the motherboard can provide, and thus dedicated connectors have been introduced to attach them directly to the power supply.[5]
- Connectors for hard drives, typically SATA only. Disk drives also connect to the power supply.

3. Other Cards

When computers communicate with people or real world, these need some devices to connect each other. Some cards are necessarily responsible for this, such as Display Card, Voice Card, Modem Card, etc.

A Video Card is a discrete dedicated circuit board, silicon chip and necessary cooling that provides 2D, 3D and sometimes even general purpose graphics processing calculations for a computer. Alternate terms include *graphics card*, *display adapter*, *video adapter*, *video board* and almost any combination of the words in these terms.

Voice card is responsible for transform digital signal to analog signal, in order to play voice through speakers or headphones.

A modem (the term is concatenated by Modulator and Demodulator) modulates outgoing digital signals from a computer or other digital device to analog signals for a conventional copper twisted pair telephone line and demodulates the incoming analog signal and converts it to a digital signal for the digital device. Currently, there are so many different modem card, such as 3G card and cable card.

4. Memory

Computer memory store massive 2 state data, 0 and 1. To choose 0 and 1 as computer's status is most electrical devices have 2 different situations, "on" and "off". If the status is "On", 1 can be stored. If the status is "Off", 0 can be stored.

All data or information (text, video, sound, and pictures), which is stored into a computer, are all presented as many 1s and 0s. So when people consider storage capacity, those following units are commonly used:

8 bits(b)=1 byte(B)
1024 bytes= 1 Kilobyte(KB)
1024 KB= 1 Megabyte(MB)
1024 MB= 1 Gigabyte(GB)
1024 GB= 1 Terabyte(TB)
1024 TB= 1 Petabyte(PB)

In computing, memory refers to the computer hardware devices used to store information for immediate use and wait for CPU to process and sometimes it is synonymous with the term "primary storage". Computer memory operates at a high speed, for example, random-access memory (RAM), as a distinction from storage that provides slow-to-access program and data storage but offers higher capacities. If needed, contents of the computer memory can be transferred to secondary storage, through a memory management technique called "virtual memory".

For computer memory, there are 2 parts. One is RAM(random access memory), the other one is ROM(read only memory). ROM is "built-in" computer memory containing data that normally can only be read, not written to. ROM contains the programming that allows users' computer to be "booted up" or regenerated each time they turn it on. Unlike a computer's random access memory (RAM), the data in ROM cannot be lost when the computer power is reboot. The ROM is sustained by a small long-life battery in computer. Therefore, if users ever do the hardware setup procedure with their computers, they effectively will be writing to ROM. Also, ROM should be seen as inner devices.

RAM (random access memory) is the place in a computer where the operating system (see Fig 1.15), application programs, and data in current use are kept so that they can be quickly reached by the computer's processor. RAM is much faster to read from and write to than the other kinds of storage in a computer, the hard disk, floppy disk, and CD-ROM. However, the data in RAM stays there only as long as people's computer is running. When users turn the computer off, RAM loses its data. When they turn computer on again, their operating system and other files are once again loaded into RAM, usually from user's hard disk. And RAM is considered as outer devices.

Fig 1.15   The image of RAM

5. I/O devices

(1) Input Devices

An input device is a peripheral (piece of computer hardware equipment), which used to transfer data and control signals to an information processing system such as a computer or information appliance. Examples of input devices include keyboards, mouse, scanners, digital cameras and joysticks.

A **keyboard** like Fig 1.16 is a human interface device which is represented as a layout of buttons. Each button, or key, can be used to either input a linguistic character to a computer, or to invoke a particular function of the computer(for example F1 to F12, Wake up, PrintScreen or etc.). They act as the main text entry interface for most users. It is typewriter like device composed of a matrix of switches.

Fig 1.16 The image of standard keyboard

A **computer mouse** is a pointing device that detects two-dimensional motion relative to a surface. This motion is typically translated into the motion of a pointer on a display, which allows a smooth control of the graphical user interface. Physically, a mouse consists of an object held in one's hand, with one or more buttons.

Fig 1.17 The image of mouse

Pointing devices are the most commonly used input devices today. A pointing device is any human interface device that allows a user to input spatial data into a computer. In the case of mouse and touchpads, this is usually achieved by detecting movement across a physical surface. Analog devices, such as 3D mice, joysticks, or pointing sticks, function by reporting their angle of deflection. Movements of the pointing device are echoed on the screen by movements of the pointer, creating a simple, intuitive way to navigate a computer's graphical user interface (GUI).

(2) Output Devices

An output device is any piece of computer hardware items used to communicate the results of data processing carried out by an information processing system (such as a computer) which converts the electronically generated information into human-readable form (see Fig 1.18).

A **display device** is an output device that visually conveys text, graphics, and video information. Information shown on a display device is called soft copy because the information exists electronically and is displayed for a temporary period of time. Display devices include CRT monitors, LCD monitors and displays, gas plasma monitors, and televisions.

Fig 1.18   Output Devices Monitor and Printer

In computers, a **printer** is a device that accepts text and graphic output in a digital signal from a computer and transfers to the real text or picture to paper, usually to some standard size sheets of paper. Printers are sometimes chosen by customers with their requirements and are purchased separately. Printers vary in size, speed, sophistication, and cost. In general, more expensive printers are used for higher-resolution color printing. By their performance, printers can be classified as stylus printer, ink printer, and laser printer.

A **digital projector**(see Fig 1.19), also called a digital projection display system, is a specialized computer display that projects an enlarged image on a movie screen. Such devices are commonly used in presentations or home theater.

There are two main types of digital projection display systems. The older, less expensive type employs three transparent liquid-crystal-display (LCD) panels, one for each of the primary colors (red, green, and blue). A newer, more expensive scheme is known as Digital Light Processing (DLP), a proprietary technology developed by Texas Instruments.

Fig 1.19   The Image of Projector

6. Storage

Except memory, in a computer system, there are so many different types of storage devices. Flash disk (USB disk), CD, and portable disks are all commonly used in computer fields (see Fig 1.20 & 1.21).

Nowadays, **Flash disk** is the most commonly used portable storage. The storage capacity of flash disk is 64KB, 128KB, or even more. People carry them very easily and also have enough storage for normal files transfer from one computer to another.

Fig 1.20　Normal storage (Flash disk and Portable disk)

Portable disk normally use for store software and documents. Users can input large-scale documents into these storage devices. Compare with USB-disk, the portable disk has bigger capacity and better performance. Using portable disk, users can move files with higher speed than the flash disk. Data in the portable disk can be transferring to a computer for any time users' want.

CD (Compact Disc) is a data storage format in the digital optical disc. This format was originally developed to store and play only sound recordings, but was later adapted for storage of data (CD-ROM). Standard CDs have a diameter of 120 millimeters (4.7 in) and can hold up to about 80 minutes of uncompressed audio or about 700 MiB of data.

Fig 1.21　The Image of Compact Disc

1.3.2　Software

Computer software, or simply **Software**, is that part of a computer system that consists of encoded information or computer instructions, in contrast to the physical hardware from which the system is built. The term is roughly synonymous with computer program, but is more generic in scope.

Base on the usage of computer system, software are divided into 3 catalogues: system software, programming software and application software.

1. System software

System software is computer software designed to provide services to other software or set the communication between human and computers. Examples of system software include operating systems, computational science software, game engines, industrial automation, and software as a service application.

The operating system (common examples are Microsoft Windows series, Mac OS X, Linux, and Unix), allows the parts of a computer to work together by performing tasks like transferring data between memory and disks or rendering output onto a display device in an understandable style-graphic. It provides a platform (hardware abstraction layer) to run high-level system software and application software.

2. Programming Software

A programming tool (or software development tool) is a type of computer program that software developers use to create, debug, maintain, or otherwise support other programs and applications. Programming Software usually refers to relatively simple programs, that can be combined together to accomplish a task. The ability to use a variety of tools productively is one distinctive characteristic of a skilled software engineer.

The most basic tools are an Integrated Development Environment (IDE) and a compiler or interpreter, which are used ubiquitously and continuously. Other tools are used more or less depending on the language, development methodology, and individual engineer, and are often used for a discrete task, like a debugger or profiler. IDE also may be discrete programs, executed separately – often from the command line – or may be parts of a single large program. In many cases, particularly for simpler use, simple and special techniques are used instead of a tool, such as print debugging instead of using a debugger, manual timing (of overall program or section of code) instead of a profiler, or tracking bugs in a text file or spreadsheet instead of a bug tracking system.

The distinction between tools and applications usually is not very clear. For example, developers use simple databases (such as a file containing a list of important values) all the time as tools. However a full-function database is usually thought of as an application or software in its own right.

3. Application Software

An application program (app or application for short) is a computer program designed to perform a group of coordinated functions, tasks, or activities for the benefit of the user. Mostly common used software are all application software, such as word processor (Microsoft Word or WPS), a spreadsheet (Excel), an accounting application (well-known as Enterprise Resources Planning or called ERP), a web browser (Internet Explorer or called as IE), a media player

(Windows Media Player), an aeronautical flight simulator, a console game or a photo editor (Photoshop). The collective name of application software refers to all applications collectively. This contrasts with system software, which is mainly involved with running the computer.

Applications may be packed with the computer and its system software or published separately, and may be coded as proprietary, open-source or university projects.

1.4　Reference

[1]　"November 2014", Top 500, Retrieved 17 January 2015.
[2]　"The List: November 2015", Top 500, Retrieved 24 January 2016.
[3]　Lemke, Tim, "NSA Breaks Ground on Massive Computing Center", Retrieved 11 December 2013.
[4]　"CPU Socket Types Explained: From Socket 5 To BGA [MakeUseOf Explains]", Retrieved 12 April 2015.
[5]　"Pinout of the PCI-Express Power Connector", techPowerUp, Retrieved 2 October 2013.

1.5　English-Chinese Key Terms

Analog device　模拟设备
Byte　字节
BIOS　基本输入输出系统
Hexadecimal　十六进制
Input　输入
Desktop　台式机
Handheld computer　掌上电脑
Tablet computer　平板电脑
Key　按键
Memory　内存
Personal computer　个人电脑
Modem　调制解调器
ALU (arithmetic logic unit)　算术逻辑单元
ASCII　美国信息交换标准代码
IDE (Integrated Development Environment)　集成开发环境
CPU (Central Processing Unit)　中央处理单元

Bit　位
Binary　二进制
Octal　八进制
Computer　计算机
Output　输出
Display device　显示器
Server　服务器
Keyboard　键盘
Mainframe computer　大型机
Slots　插槽
Motherboard (mainboard)　主板
Notebook　笔记本电脑

Chapter 2　Operating System
操作系统

In this chapter, we will introduce the types, functions, and common features of an operating system. We will also see how the latest Microsoft Operating System (win 10) is working, including the management of files, Control Panel and some common use accessories.

2.1　An Introduction of Operating System

An **operating system** (**OS**) is system software that manages computer hardware and software resources and provides common services for computer programs. Also, OS is a program which presents as an intermediate between users and the computer hardware. It provides a user-friendly environment in which a user may easily develop and execute programs. Otherwise, hardware knowledge would be mandatory for computer programming. Therefore, computer users can think that an OS is a bridge from the complexity of hardware to understandable operation.

The operating system is a component of the system software in a computer system. Application programs usually require an operating system to function.

2.1.1　What is an Operating System

The Operating System manages these resources and allocates them to specific programs and users. With the management of the OS, a programmer is rid of difficult hardware considerations. Usually, an OS provides services for following function:
- Processor Management
- Memory Management
- File Management
- Device Management
- Concurrency Control

Another aspect for the usage of OS is used as a predefined library for hardware-software interaction. It is the reason why system programs apply to the installed OS since they cannot reach hardware directly.

2.1.2　Operating System Classification

Since operating system involve into computer field, massive types of OS have been used in different way. Users define OS catalog as different contents.

1. Base on their work

Single- and multi-tasking: A single-tasking system can only run one program at a time, while a multi-tasking operating system allows more than one program to be running in concurrency. This is achieved by time-sharing, dividing the available processor time between multiple processes that are each interrupted repeatedly in time slices by a task-scheduling subsystem of the operating system. Multi-tasking may be characterized in preemptive and co-operative types. In preemptive multitasking, the operating system slices the CPU time and dedicates a slot to each of the programs (Fig 2.1).

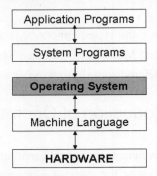

Fig 2.1 OS: the interaction between hardware and software

Single- and multi-user: Single-user operating systems have no facilities to distinguish users, but may allow multiple programs to run in a queue. A multi-user operating system extends the basic concept of multi-tasking with facilities that identify processes and resources, such as disk space, belonging to multiple users, and the system permits multiple users to interact with the system at the same time. Time-sharing operating systems schedule tasks for efficient use of the system and may also include accounting software for cost allocation of processor time, mass storage, printing, and other resources to multiple users.

2. Base on their usage

Super-computing is primarily scientific computing, usually modeling real systems in nature. Render farms are collections of computers that work together to render animations and special effects. Work that previously required supercomputers could be done with the equivalent of a render farm. Such computers are found in public research laboratories, Universities, Weather Forecasting laboratories, Defense and Energy Agencies, etc.

Mainframes used to be the primary form of computer. Mainframes are large centralized computers. At one time, they provided the bulk of business computing through time-sharing. Mainframes and mainframe replacements (powerful computers or clusters of computers) are still useful for some large-scale tasks, such as centralized billing systems, inventory systems, database operations, etc. When mainframes were in widespread use, there was also a class of computers known as minicomputers that were smaller, less expensive versions of mainframes for businesses that could not afford mainframes.

Servers are computers or groups of computers used for Internet serving, intranet serving, print serving, file serving and/or application serving. Clustered Servers are sometimes used to replace mainframes.

Desktop operating Systems are used on standalone personal computers.

Workstations are more powerful versions of personal computers. Often only one person uses a particular workstation that run a more powerful version of a desktop operating system. They usually have software associated with larger computer systems through a LAN network.

Handheld Operating Systems are much smaller and less capable than desktop operating systems, so that they can fit into the limited memory of handheld devices. Barcode scanners, PDA's, are examples of such systems. Currently, the PDA world is witnessing an operating system battle between several players (Microsoft Windows, IOS, etc.)

Real Time Operating Systems (RTOS) are designed to respond to events that happen in real time. Computers using such operating systems may run ongoing processes in a factory, emergency room systems, air traffic control systems or power stations. The operating systems are classified according to the response time they need to deal with: seconds, milliseconds, micro-seconds. They are also classified according to whether or not they involve systems where failure can result in loss of life. As in the case of supercomputers, there are no such systems in Lebanon today. However, given the way the technology is growing, it may be possible to use them in the future.

Embedded Systems are combinations of processors and special software that are inside another device, such as contents switches or Network Attached Storage devices.

Smart Card Operating Systems are the smallest Operating Systems of all. Some handle only a single function, such as electronic payments, others handle multiple functions. Often these OS are proprietary systems but we are seeing more and more smart cards that are Java oriented.

Specialized Operating Systems, like Database Computers are dedicated high performance data warehousing servers.

3. Historical Operation Systems

- MS-DOS (Microsoft Disk Operating System)

Originally developed by Microsoft for IBM, MS-DOS (Fig 2.2) was the standard operating system for IBM-compatible personal computers. The initial versions of DOS were very simple and resembled another operating system called CP/M. Subsequent versions have become increasingly sophisticated as they incorporated features of minicomputer operating systems.

Fig 2.2　Image of MS-DOS

- Windows 1.0 – 2.0 (1985－1992)

Introduced in 1985, Microsoft Windows 1.0 was published that represented a fundamental aspect of the operating system. Instead of typing MS-DOS commands, windows 1.0 allowed users to point and click to access the windows.

In 1987 Microsoft released Windows 2.0, which was designed for the designed for the Intel 286 processor. This version added desktop icons, keyboard shortcuts and improved graphics support.

- Windows 3.0 – 3.1 (1990–1994)

Windows 3.0 was released in May, 1990 offering better icons, performance and advanced graphics with 16 colors designed for Intel 386 processors. This version is the first release that provides the standard "look and feel" of Microsoft Windows for many years to come. Windows 3.0 included Program Manager, File Manager and Print Manager and games (Hearts, Minesweeper and Solitaire). Microsoft released Windows 3.1 in 1992.

- Windows 95 (August 1995)

Fig 2.3　Image of Windows 95

As its named, Windows 95 (Fig 2.3) was released in 1995 and was a major upgrade to the Windows operating system. This OS was a significant advancement over its former versions. In addition to presenting a new user interface, Windows 95 also includes a number of important internal improvements. For example, it supports 32-bit applications, which means that applications written specifically for this operating system should run much faster.

Although Windows 95 can run older Windows and DOS applications, it has essentially removed DOS as the underlying platform. This has meant removal of many of the old DOS limitations, such as 640K of main memory and 8-character filenames. Other important features in this operating system are the ability to automatically detect and configure installed hardware (plug and play).

- Windows NT 3.1 – 4.0 (1993－1996)

A version of the Windows operating system. Windows NT (New Technology) is a 32-bit operating system that supports preemptive multitasking. There are actually two versions of Windows

NT: Windows NT Server, designed to act as a server in networks, and Windows NT Workstation for stand-alone or client workstations.

- Windows 2000 (February 2000)

Often abbreviated as "W2K", Windows 2000 (like Fig 2.4) is an operating system for business desktop and laptop systems to run software applications, connect to Internet and intranet sites, and access files, printers, and network resources. Microsoft released four versions of Windows 2000: Professional (for business desktop and laptop systems), Server (both a Web server and an office server), Advanced Server (for line-of-business applications) and Datacenter Server (for high-traffic computer networks).

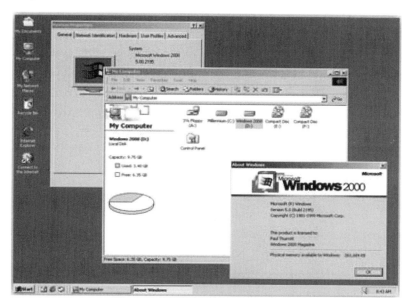

Fig 2.4　Image of Windows 2000

- Windows XP (October 2001)

Windows XP, was released in 2001. Along with a redesigned look and feel to the user interface, the new operating system is built on the Windows 2000 kernel, giving the user a more stable and reliable environment than previous versions of Windows. Windows XP comes in two versions, Home and Professional. Because of the stability, reliability, and common usage, Windows XP perhaps is the longest usage operating system than ever.

- Windows 7 (October 2009)

Windows 7 was released by Microsoft on October 22, 2009, as the latest in the 25-year-old line of Windows operating systems and as the successor to Windows Vista (which itself had followed Windows XP). Windows 7 was released in conjunction with Windows Server 2008 R2, Windows 7's server counterpart. Enhancements and new features in Windows 7 include multi-touch support, Internet Explorer 8, improved performance and start-up time, Aero Snap, Aero Shake, support for virtual hard disks, a new and improved Windows Media Center, and improved security.

- Windows 8

Windows 8 was released on August 1, 2012 and is a completely redesigned operating system that's been developed from the ground up with touchscreen use in mind as well as near-instant-on capabilities that enable a Windows 8 PC to load and start up in a matter of seconds rather than in minutes.

Windows 8 will replace the more traditional Microsoft Windows OS look and feel with a new "Metro" design system interface that first debuted in the Windows Phone 7 mobile operating system. The Metro user interface primarily consists of a "Start screen" made up of "Live Tiles" which are links to applications and features that are dynamic and update in real time.

- Windows 10

Windows 10 is Microsoft's Windows successor to Windows 8. Windows 10 debuted on July 29, 2015, following a "technical preview" beta release of the new operating system that arrived in Fall 2014 and a "consumer preview" beta in early 2015. Microsoft claims Windows 10 features fast start up and resume, built-in security and the return of the Start Menu in an expanded form. This version of Windows will also feature Microsoft Edge, Microsoft's new browser. Any qualified device (such as tablets, PCs, smartphones and Xbox consoles) can upgrade to Windows 10, including those with pirated copies of Windows.

- UNIX

Unix (often spelled "UNIX," especially as an official trademark) is an operating system that originated at Bell Labs in 1969 as an interactive time-sharing system. Ken Thompson and Dennis Ritchie are considered the inventors of UNIX. The name (pronounced YEW-nihks) was a pun based on an earlier system, Multics. In 1974, UNIX became the first operating system written in the C Programming language. UNIX has evolved as a kind of large freeware product, with many extensions and new ideas provided in a variety of versions of UNIX by different companies, universities, and individuals.

UNIX operating systems are used in widely-sold workstation products from Sun Microsystems, Silicon Graphics, IBM, and a number of other companies. The Unix environment and the client/server program model were important elements in the development of the Internet and the reshaping of computing as centered in networks rather than in individual computers. Linux, a Unix derivative available in both "free software" and commercial versions, is increasing in popularity as an alternative to proprietary operating systems.

- Linux

Linux (often pronounced LIH-nuhks with a short "i") is a Unix-like operating system that was designed to provide personal computer users a free or very low-cost operating system comparable to traditional and usually more expensive Unix systems. Linux has a reputation as a very efficient and fast-performing system. Linux's kernel (the central part of the operating system) was developed by Linus Torvalds at the University of Helsinki in Finland. To complete the operating system, Torvalds and other team members made use of system components developed by members of the Free Software Foundation for the GNU Project.

Linux is a remarkably complete operating system, including a graphical user interface, an X Window System, TCP/IP, the Emacs editor, and other components usually found in a comprehensive Unix system. Although copyrights are held by various creators of Linux's components, Linux is distributed using the Free Software Foundation's copyleft stipulations that mean any modified version that is redistributed must in turn be freely available.

Unlike Windows and other proprietary systems, Linux is publicly open and extendible by contributors. Because it conforms to the Portable Operating System Interface standard user and programming interfaces, developers can write programs that can be ported to other operating systems. Linux comes in versions for all the major microprocessor platforms including the Intel, PowerPC, Sparc, and Alpha platforms. It's also available on IBM's S/390. Linux is distributed commercially by a number of companies. A magazine, Linux Journal, is published as well as a number of books and pocket references. Linux is sometimes suggested as a possible publicly-developed alternative to the desktop predominance of Microsoft Windows. Although Linux is popular among users already familiar with Unix, it remains far behind Windows in numbers of users. However, its use in the business enterprise is growing.

- Embeded Operation System

Android is a mobile operating system (OS) currently developed by Google, based on the Linux kernel and designed primarily for touchscreen mobile devices such as smartphones and tablets. Android's user interface is mainly based on direct manipulation, using touch gestures that loosely correspond to real-world actions, such as swiping, tapping and pinching, to manipulate on-screen objects, along with a virtual keyboard for text input. In addition to touchscreen devices, Google has further developed Android TV for televisions, Android Auto for cars, and Android Wear for wrist watches, each with a specialized user interface. Variants of Android are also used on notebooks, game consoles, digital cameras, and other electronics.

iOS (originally iPhone OS) is also a mobile operating system created and developed by Apple Inc. and distributed exclusively for Apple hardware. It is the operating system that presently powers many of the company's mobile devices, including the iPhone, iPad, and iPod touch. It is the second most popular mobile operating system in the world by sales, after Android. iPad tablets are also the second most popular, by sales, against Android since 2013, when Android tablet sales increased by 127%.

2.1.3 Management of Files

1. Icons

In computing, an icon is displayed on a computer screen in order to help the user navigate a computer system or mobile device. The icon itself is a quickly comprehensible symbol of a software tool, function, or a data file, accessible on the system and is more like a path than a detailed illustration of the actual entity it represents. It can serve as an electronic hyperlink or file shortcut to access the program or data. The user can activate an icon using a mouse, pointer, finger, or recently voice commands. Their placement on the screen, also in relation to other icons, may provide further information to the user about their usage. In activating an icon, the user can move directly into and out of the identified function without knowing anything further about the location or requirements of the file or code.

Desktop icons for file/data transfer, clock/awaiting, and running a program.

Icons as parts of the graphical user interface of the computer system, in conjunction with windows, menus and a pointing device (mouse), belong to the much larger topic of the history of the graphical user interface that has largely supplanted the text-based interface for casual use.

Now, we are introducing 2 commonly used icons.

- "This PC": is use for easily access basic functions for user to manage their files, folders, or computer disks. This is the main access for visit documents, camera, scanner, printer, other devices, and other information. "This PC" is one of the most important functions in computers for users' control (see Fig 2.5).
- "Recycle Bin": The Recycle Bin in used by Windows computers to store deleted items. For example, "If users want to delete items in Windows, drag them to the Recycle Bin.". It temporarily stores files and folders before they are permanently deleted. People can open the Recycle Bin by double-clicking the icon on the Windows desktop (see Fig 2.6).

Fig 2.5   Image of This PC   Fig 2.6   Image of Recycle bin

2. Icons' management

On the desktop, there are different types of icons (see Fig 2.7).

A **shortcut icon**, with little arrow in the left corner, provides alternative access to programs and documents. Usually the size of one shortcut icon is less than 1KB, which is a path between icon and the source program and can minimize occupation of C: disk. Also deleting a shortcut does not delete the item itself.

A **taskbar** is an element of a graphical user interface which has various purposes (see Fig 2.8). It typically shows which programs or applications are running on the device, as well as provide links or shortcuts to other programs or places, such as a start menu, notification area, and clock.

- The Start button, a button that invokes the Start menu. It appears in Windows 9x, Windows NT 4.0 and all its successors, except Windows 8 and Windows Server 2012. The Start button presents an opening window in several parts (see Fig 2.9).
  - ➢ Commonly used area: this part usually shows the program that users use often, for example, Internet Explorer.
  - ➢ Program area: it can display the entire program in the computer. The most frequently used programs will show and others can be shown in clicking All Programs button.
  - ➢ The right area: it shows short path with computer's functions. For example, Search and Gmail.

Chapter 2   Operating System
操作系统

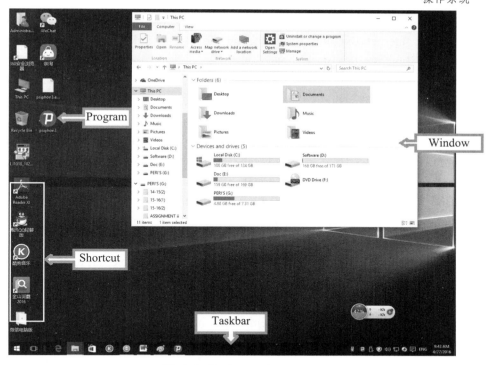

Fig 2.7   Desktop and Icons

Fig 2.8   Taskbar

Fig 2.9   Start button

- The Quick Launch bar(or Program buttons), introduced on Windows 95 and Windows NT 4.0 through the Windows Desktop Update for Internet Explorer 4 and bundled with

Windows 98, contains shortcuts to applications. Windows provides default entries, such as Launch Internet Explorer Browser, and the user or third-party software may add any further shortcuts that they choose. A single click on the application's icon in this area launches the application. This section may not always be present, for example it is turned off by default in Windows XP and Windows 7.

- The notification area is presenting the taskbar that displays icons for system and program features that have no presence on the desktop or running in the backstage during the operation system started as well as the time and the volume icon. It contains mainly icons that show status information, though some programs, such as Firewall and anti-virus software, use it for minimized windows. By default, this is located in the bottom-right of the primary monitor, or at the bottom of the taskbar if docked vertically. The clock appears here, and applications can put icons in the notification area to indicate the status of an operation or to notify the user about an event. For example, an application might put a printer icon in the status area to show that a print job is under way, or a display driver application may provide quick access to various screen resolutions. The notification area is commonly referred to as the system tray, which Microsoft states is wrong, although the term is sometimes used in Microsoft documentation, articles, software descriptions, and even applications from Microsoft such as Bing Desktop. The notification area is also referred to as the status area by Microsoft.

2. Icons' management

Because of the convenience of desktop icon, users would like use them directly. Usually, user should arrange icons by their behaviors.

To arrange icons, users right click mouse button in any empty space of the desktop and select Sort by. And user can choose the arrangement rules by different requirement (see Fig 2.10).

Fig 2.10 Arrange icons

- Name: arrange icons by their name with alphabet order.
- Size: arrange icons by files' size.

- Item type: arrange icons files' type.
- Date modified: arrange icons by the latest time of file modification.

3. New function in Win 10 —— Cortana

Cortana is easily one of the coolest new features of Windows 10. Users are probably already familiar with personal assistants like Google Now and Siri, but now they have one built right into their desktop. Cortana can provide so many functions, such as manage users' schedule, how to get information, and even run a few Google commands.

- Using Cortana: Voice vs Typing

Fig 2.11 Cortana Can Do Everything in Windows 10

There are two ways to use Cortana: use voice commands, or type out commands in the Start Menu. If choose the former, the user may want to enable the "Hey Cortana" feature. With this on, the user can say "Hey Cortana" out loud to trigger voice commands without pressing a button.

Search for "Cortana settings" in Start Menu.

Enable the toggle under "Hey Cortana".

(Optional) Under "Respond best", users can choose "to me" to tailor Cortana by their voices. They'll need to perform a couple quick exercises to teach Cortana users' voice. Otherwise, Cortana will work for anyone.

In general, users can use Cortana to search the web for just about anything. Out of the box, Cortana uses Bing for web results, so the answers are a bit limited. However, there are still plenty of answers users can get from Bing even without changing search engines.

4. Set wallpaper

Microsoft went to a great deal of effort to create the Desktop wallpaper that comes standard with Windows 10. It even produced a video showing its creation. But what if, users would like to change their wall paper frequently or change the wrong wallpaper by accidentally. Fear not, the same customizations are possible in Windows 10 that have been a staple of earlier Windows versions.

To get started, simply right-click on the desktop. Alternatively, users could go to Settings (via the new Action Center's "All settings" button or by typing Settings in Cortana's search box) (see Fig 2.12).

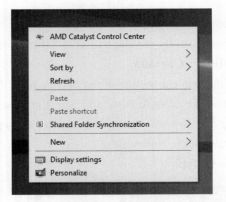

Fig 2.12　Set wallpaper

Choose Personalization (or Personalize, if users are going from the right-click on the desktop). Here's the window that will appear like Fig 2.13.

Fig 2.13　Personalization Settings in Windows 10

Note it shows users a preview of what their desktop looks like as they try different options in the Settings dialog. They will have a choice between using a picture, a solid color, or a slideshow as the desktop background, aka wallpaper. See Fig 12.14.

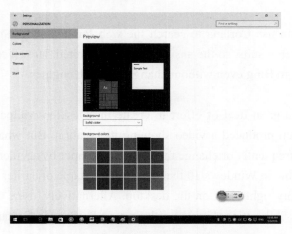

Fig 2.14　Solid Background Wallpaper in Windows 10

A preset sampling of five pictures appears under "Choose your picture", but users are not limited to those. Far from it, people can tap Browse to choose any image file on their hard drive that's in JPG, JPEG, BMP, DIB, PNG, JFIF, JPE, GIF, TIF, TIFF, or WDP format. After people have selected the wallpaper of their favorites, they need to decide how to fit it on the screen, since not every image has the same aspect ratio as the monitor. Their fit choices are Fill, Fit, Stretch, Tile, Center, and Span. That last lets users spread an image across multiple monitors, an option that Windows 8 introduced.

Windows 10 also provides the same old picture on people's desktop all the time is too monotonous; the Personalization settings also give users the option of having their wallpaper play a slow slideshow. With this option, choose a folder containing background images of the same formats listed above, rather than specific images. People can set the photos to switch at intervals of one minute, 10 minutes, 30 minutes, one hour, six hours, or a day.

Another, simpler option for changing the desktop wallpaper that bypasses the Settings dialog will be known to many users. While viewing any image in File Explorer, users can simply right-click on the image and choose "Set as desktop background".

- Themes

If users want customization that goes beyond just the background, as in past Windows versions, people have the option of implementing Themes. These are packs of backgrounds, interface colors, and even sounds. There are three standard Windows Themes available, and any that users had in their previous version of Windows remain available. To get to Themes, click on the fourth option down in the Personalization settings window, and then on Theme settings (this seems like an extra unnecessary click, but the Themes page also lets users choose sound, desktop icon, and mouse pointer settings). It helps users to the old-style Control Panel for Personalization Windows 10 duplicates some features like this (see Fig 2.15).

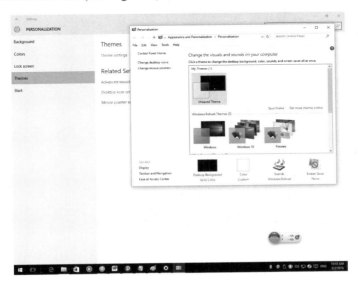

Fig 2.15 Themes in Windows 10

Users can download many more themes on Microsoft's website. There are loads of choices there, including landscapes, animals, flowers, art, holidays, and community showcases. Microsoft has unfortunately retired one of favorite theme options—the dynamic theme, which used RSS to download ever-changing backgrounds. So users will just have to select photos on their own or choose one of the many downloadable themes.

2.1.4 Work with Control Panel

Despite first appearances, Windows 10 hasn't cancelled the Control Panel. Although users are strongly pushed towards the new – and admittedly improved – settings dialogue first seen in Windows 8, the old Control Panel still lurks behind the scenes, giving users access to just about every system setting imaginable (see Fig 2.16).

And while a fair bit of its functionality has now been incorporated into Settings, there are still plenty of times when users will need to access the Control Panel to tweak certain parts of people's PC.

In some cases, direct links to some Control Panel applets can be found in Settings itself, look for references to "advanced settings" to access them. However, in this tutorial we'll focus on revealing key settings that are hidden behind the scenes, read on to find out more about them, plus gain that all-important access.

Fig 2.16 Open Setting

To open Control Panel, users can search Control Panel in Cortana's search box or open Setting option window and type Control Panel in the Find a setting box.

Chapter 2　Operating System
操作系统

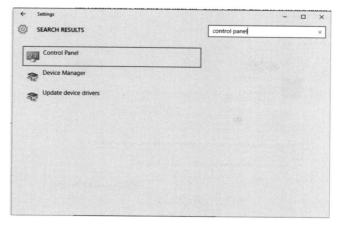

Fig 2.17　Search Control Panel

In this case, the classic control panel icon can be shown in the display window. When users single click the icon, Control Panel window shows, like Fig 2.18.

Fig 2.18　Control Panel Display

As the figure shows above, there are so many functions which can be used in the Control Panel. Therefore, how to use the Control Panel is the core question for normal users. The Control Panel itself is really just a collection of individual Control Panel applets so to use the Control Panel really means to use an individual applet to change some part of how Windows works.

Here are a few of the thousands of individual changes that are possible from within Control Panel:

- Change Your Password

Click on the User Accounts link. In the Make changes to user account area of the User Accounts window, click the Change your password link. In the first text box, enter current password. In the next two text boxes, enter the password to start using. Entering the password twice helps to make sure that users typed new password correctly. In the final text box, users are asked to Type a password hint (Fig 2.19).

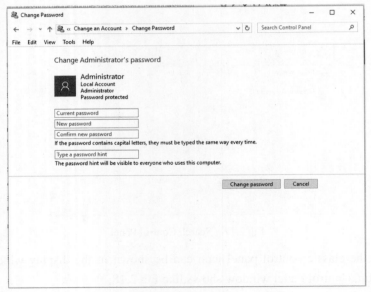

Fig 2.19　　Change Your Password

- Change Another User's Password

On Windows 10, touch or click on the User Accounts link. There are several links down on the make changes to user account area of the User Accounts window, touch or click on Manage another account. Touch or click on the "user you would like to change" the password for.

- Create a Password for Your Account

Similar with other operation, users should click on the User Accounts link. In the Make changes to user account area of the User Accounts window; click the Create a password for users' account link. In the first two text boxes, enter the password users would like to start using. Entering the password twice helps to make sure that users typed their new password correctly. In the final text box, users are asked to Type a password hint.

- Adjust the Date and Time

If the date or time is incorrect for users' location, select the Date and Time Settings link. The Settings screen opens. Users can change directly (see Fig 2.20).

Also, people can change Date and time by Select Time & Language windows.

Windows 10 determines the correct time and date from the Internet, and users' computer should show the right time and date. If it doesn't, turn off the Set Time automatically option and select the "Change" button. Users see the Change Date and Time screen. Select the correct date and time in this screen.

Change the time by using the little triangles that point up (later) or down (earlier) or by entering the specific hours and minutes. Select Change to keep change or Cancel to ignore change.

Back in the Date & Time window, select users Time Zone from the drop-down list, if necessary. Turn the Adjust for Daylight Saving Time option on or off as appropriate, like Fig 2.21.

Chapter 2　Operating System
操作系统　37

Fig 2.20　Adjust the Date and Time

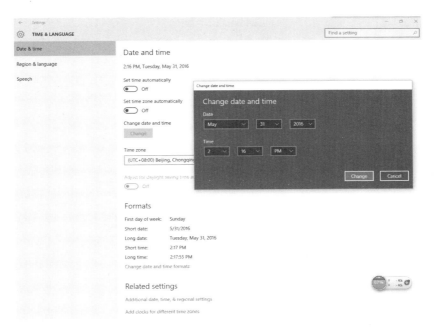

Fig 2.21　Time & Language windows

　　To change the format with which the time and date are displayed in the lower-right corner of the screen, select the Change Date and Time Formats link in the Date & Time window. Then, in the Change Date and Time Formats window, choose how users want the date and time to be displayed on computer screen.

- Start Device Manager

Tap or click on the Hardware and Sound link.

From this Control Panel screen, look for and choose Device Manager. In Windows 10, check under the Devices and Printers heading.

- Install Windows Updates

First, tap or click on the "Start" menu, followed by "Settings". Once there, choose "Update & security", followed by "Windows Update" on the left. Check for new Windows 10 updates by tapping or clicking on the "Check for updates" button. In Windows 10, downloading and installing updates is automatic and will happen immediately after checking or, with some updates, at a time when users are not using their computers.

- Hide Hidden Files

Click on the "Folder Explorer Options link". Click on the "View tab" on the Folder Explorer Options window. In the Advanced settings: area, locate the Hidden files and folders category. Choose the "Don't show hidden files, folders, or drives radio" button under the Hidden files and folders category. Click "OK" at the bottom of the Folder Options window. Close the Control Panel window that's still open. From now on, no files or folders with the hidden attribute turned on will display in any folder view or search (see Fig 2.22).

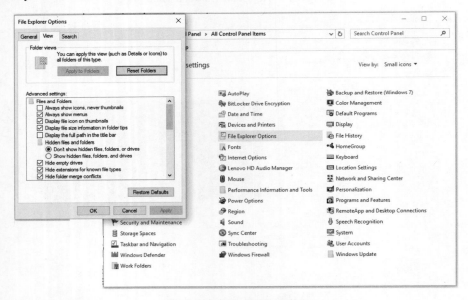

Fig 2.22 Hide Hidden Files

- Change the Default Program for a File Extension

Similar steps with Hide Hidden Files with open File Explorer Options window (Shown on Fig 2.22), users can find an option with "Hide extensions for known file types". If choose this option, click "Apply" or "OK" button and all files' extension will show. And then, users could Right click mouse button, choose Rename to modify files' extension.

- Reinstall a Program

Open Settings, and click/tap on the System icon. In System settings, click/tap on "App & features" on the left side (see Fig 2.23&2.24).

Chapter 2　Operating System

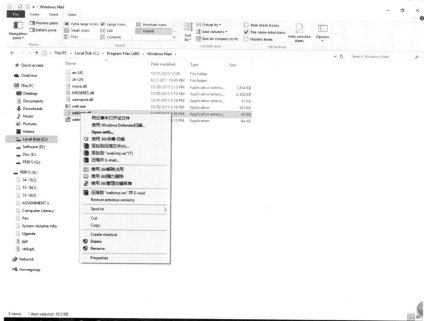

Fig 2.23　Change files' extension

Fig 2.24　App & features

- Uninstall a Program

On the right side, click/tap on a Windows app or desktop app that users want to uninstall, and click/tap on the Uninstall button.

Click/tap on Uninstall to confirm.

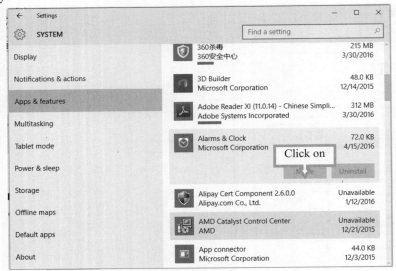

Fig 2.25　Uninstall files

2.2　Accessories

Windows accessories or called applications provide several convenient used software, which can help user do better work. Unlike Windows XP, users who were running the Windows 7 Operating System and now upgrade their Operating System to Windows 10 will be confused when they look for the accessories. The procedure for accessing the Accessories folder was different in Windows 7. The Accessories folder is accessible through the Start Menu of Windows 7. In the next operating system, due to changes of the Windows operating system interface, it became harder to find the Accessories folder. But the current Windows 10 is totally different. Users can now see a few tools listed in the All Apps section of Start Menu whereas a few tools are still in the Accessories folder. Now, if users want to access some important tools then users need to get them in the Accessories folder. The procedure in this article is very easy and simple to follow and allow accessing the tools.

Firstly, execute a hit on the Windows logo and fire up Start Menu which should be followed by a click on All Apps button. Secondly, pass through all the sections and make a halt at the "W" section of All Apps. There users can see a folder named Windows Accessories (see Fig 2.26).

A drop-down menu is also associated with the folder. Click on it and users can see all items listed henceforth.

- Character Map

Character Map enables users to view the characters that are available in a selected font. Using Character Map, users can copy individual characters or a group of characters to the Clipboard and paste them into any program that can display them.

Fig 2.26 Windows accessories

When users want to insert a special character, they should Click the Font list, and then click the font people want to use. Click the special character they want to insert into the document. Click "Select", and then click "Copy". Open their document and click the location in the document where users want the special character to appear. Click the "Edit" menu, and then click Paste (shown in the Fig 2.27).

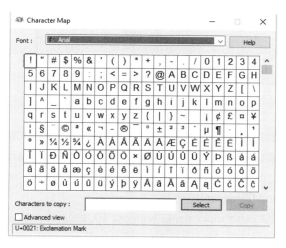

Fig 2.27 Character Map

- Math Input Panel

Windows 10 has a small application called Math Input Panel. It was active in the previous versions of Windows too. With the help of input devices like the mouse, touch screens, etc. Users create the mathematical formulas on the application and then can insert them in the documents for further use. In the Math Input Panel application, the formulas that users enter are inserted in the editable forms in the documents where people can edit the output as users edit the other types of text. This application is very useful especially when users have to create documents with a number of mathematical formulas in it.

When users open the Math Input Panel, they can view "Write Math here" in the core of the application. Write formulas or equations there. After users start inserting the formula, they will see the tools on the right side of the application's window.

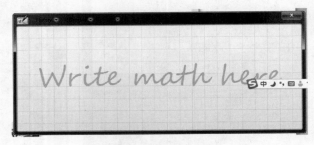

Fig 2.28 Math Input Panel

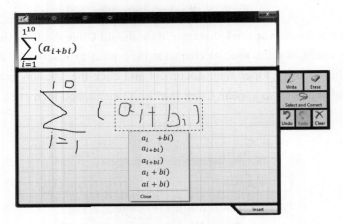

Fig 2.29 Input a formula

Tools on the right side of the math input panel application: The tools available on the right side are Write, Erase, Select and Correct, Undo, Redo and Clear. Use the Erase tool if users need to erase some part of the formula. And then click on Write to finish the erased part. Users also can use functions of Undo, Redo, and Clear tools.

To write formula on the math input panel window 10, users are done with the formula click on Insert button on the bottom right of the window. This will paste the formula automatically in the document where users wish to use it. If users see that the formula has not been copied itself in the application where their wished for, then there is nothing to worry about. The Math Input Panel saves their written formula in the clipboard by default. So use the keyboard shortcuts "Ctrl+V" in the application and users will see it inserted.

How to insert formula in Microsoft word? If in a single session, users enter various formulas, using the History menu users can get back to any one of them. When users select the History menu, choose the former formula that they wish to edit or even if people wish to insert it again in the document. As the formula becomes ready in the Math Input Panel application, edit it, if needed and select the Insert button to get it in their document.

If users need correct formulas in Math Input Panel in certain situations, Math Input Panel cannot identify users' writing clearly. And to fix it, users can use the Select and Correct tool. Click on the tool, and then select the character that people wish to correct. A drop down menu will arise exhibiting few options. See them and select the one that users intended to use there. And take up their writing again by click on the Write tool.

- Paint

Paint is a feature in Windows that people can use to create drawings on a blank drawing area or in existing pictures. Many of the tools users use in Paint are found in the ribbon, which is near the top of the Paint window(see Fig 2.30).

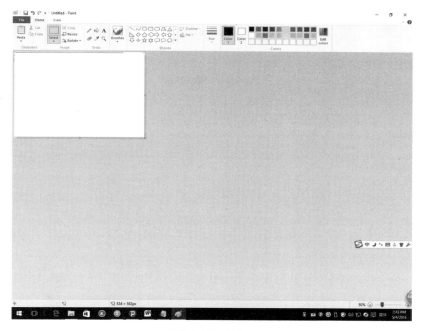

Fig 2.30   Paint

- WordPad

Microsoft WordPad is a free rich text editor included with Microsoft Windows. Although capable of doing more than Notepad, WordPad is not as advanced as Microsoft Word. However, it gives users additional features such as the capability of inserting pictures and text formatting. The picture below (Fig 2.31) shows an example of Microsoft WordPad.

Microsoft WordPad is capable of editing and saving plain-text file (.txt), Rich Text Format (.rtf), and Word for Windows 6.0 (.doc or .docx), and OpenDocument Text (.odt) format files. However, not all versions of WordPad support all above formats. Windows 95, Windows 98, Windows ME, and Windows XP does not support .doc. Windows 7 introduced the support of .odt files, so early versions of Windows do not support this format.

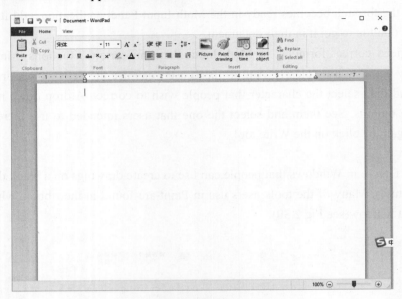

Fig 2.31   WordPad

Chapter 3　Microsoft Word 2010
文字处理软件——Word 2010

3.1　An introduction of MS Word 2010

Microsoft Office 2010 is a version of the Microsoft Office productivity suite for Microsoft Windows. Microsoft Office 2003 and the Office 2007 are very successful in people's usage. Compare with those 2 old versions, Office 2010 includes extended file format support, user interface improvements, and a changed user experience. Nowadays, the 64-bit version of Office 2010 provides in the world, however, that is not suitable for Windows XP or Windows Server 2003. It is the first version of the productivity suite to ship in both 32-bit and 64-bit versions.

On April 15, 2010, Office 2010 was released to manufacturing. The suite was subsequently made available for retail and online purchase on June 15, 2010. Office 2010 is the first version to require product activation for volume licensing editions. Unlike previous versions of the productivity suite, every application in Office 2010 features the ribbon as its user primary user interface.

The whole package of Office 2010 includes Word, Excel, PowerPoint, Access and Info Path. Each one of the package charges different functions in computer field. For example, Word is responsible for documents editing, Excel is for data analysis and diagram presenting, PowerPoint is focus on presentation or meeting situation, and Access can be used as a small-scale database.

In this chapter, Word 2010 is the main function to be introduced (see Fig 3.1).

3.1.1　Word 2010 Components

As one of the most common computer application software, Microsoft Word provides several functions for normal users. Users navigate Word 2010 by click Start button on the taskbar and use Word to type, edit, modify, convert, and other operations of document.

1. Quick Access Toolbar (QAT): Displays buttons to perform frequently used commands with a single click. Frequently used commands in Word include Save, Undo, Redo, and Print. For commands that users use frequently, they can add additional buttons to the Quick Access Toolbar.

2. Ribbon: Organizes commands on tabs, and then groups the commands by topic for performing related document tasks.

3. File Tab: Displays Microsoft Office Backstage view, which is a centralized space for all of user's file management tasks such as opening, saving, printing, publishing, or sharing a file. Tabs Display across the top of the Ribbon, and each tab relates to a type of activity; for example, laying out a page.

4. Group name: It Indicate the name of the groups of related commands on the displayed tab.

5. Dialog box launcher: It is a small icon that displays to the right of some group names on the Ribbon; it launches a dialog box.

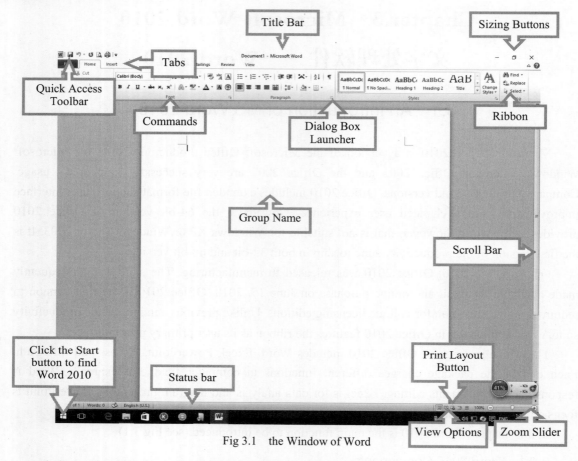

Fig 3.1   the Window of Word

6. Status bar Displays: On the left side, the page and line number, word count, and the Proof button. On the right side, displays buttons to control the look of the window.

7. Print Layout button: The default view, which displays the page borders and the document as it will appear when printed.

8. View Options: Those contain buttons for viewing the document in Print Layout, Full Screen Reading, Web Layout, Outline, or Draft views, and also display controls to Zoom Out and Zoom In.

9. Zoom Slider: The Zoom Slider increases or decreases the viewing area.

10. Vertical scroll bar: It enables users to move up and down in a document to display text that is not visible.

11. Scroll box: It provides a visual indication of displayed location in a document. Users can use the mouse to drag a document up and down to reposition the document.

12. Sizing Buttons: The buttons on the right edge of the title bar that minimize, restore or close the program.

13. Title bar: It displays the name of the document and the name of the program.

3.1.2 Word 2010 Layouts

There are several layouts can be display in Word 2010: Print Layout, Full Screen Reading, Web layout, Outline, and Draft format.

1. Print Layout view: When users start a new document, they are already in Print Layout view, which is where users do their writing. Sometimes, perhaps users inherited this document and it's not in Print Layout view or users accidentally changed the view they were in and now they want to get back to Print Layout to click "Print Layout view".

2. Full Screen Reading: is optimized for reading a document on the computer screen. In Full Screen Reading view, users also have the option of seeing the document as it would appear on a printed page. (Fig 3.2)

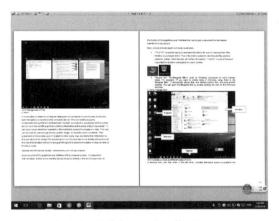

Fig 3.2 Full Screen Reading Layout

3. Web Layout: The Web Layout displays the document as if it was a web page.
4. Outline: The display of this Layout is document's outline, like Fig 3.3.

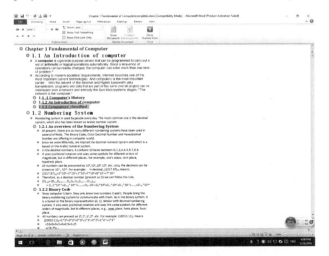

Fig 3.3 Outline Layout

5. Draft Format: allows quick text editing and formatting; headers and footers will not be shown (Fig 3.4).

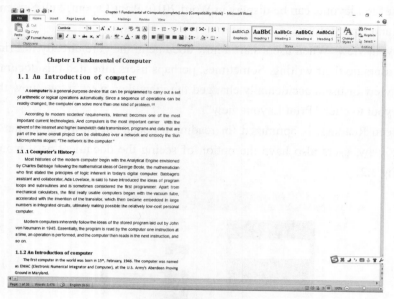

Fig 3.4   Draft format

3.2   Creating a document

Before creating a document, users should start Word software first. And there are several ways to start up (Fig 3.5).

Fig 3.5   Start Word 2010 from Start button

1. Start button
- Start Windows 10 first.
- Click "Start Button"→"All apps"→go for "M"→under "Microsoft Office", choose "Word" to start.

2. Use "Shortcut": if a "Word Shortcut" is on the desktop, users can start the software by double click shortcut icons.

3. In a folder or desktop, mouse right click, and choose "New"→"Microsoft Word Document", and then double click to start (see Fig 3.6).

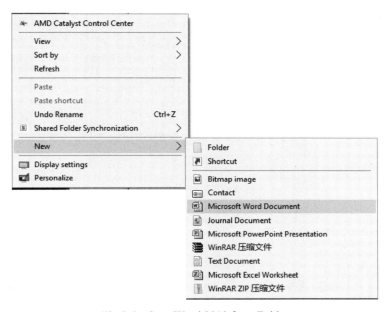

Fig 3.6　Start Word 2010 from Folder

3.2.1　Creating a new document

After starting Word, the system will open a blank document automatically and the default's name is "Document 1". This blank document can be started writing by users. After Word has already started and users are ready to begin another new document, users summon the electronic equivalent of a fresh, blank sheet of paper:

1. Click the "File" tab: The Word window changes to display the File tab menu.

2. Select the New command from the left side of the window. Word lists a slew of options for starting a new document, many of which may appear confusing to users. What users wants are the Blank Document item, which is conveniently chosen for user (Fig 3.7).

3. Click the Create button to start a new, blank document. The Create button is found on the right side of the window, beneath that obnoxiously large, blank sheet of paper. Alternately, people can simply press "Ctrl+N" with Word open to start a new, blank document. The Word window returns to normal and they see a blank page, ready for typing like Fig 3.8.

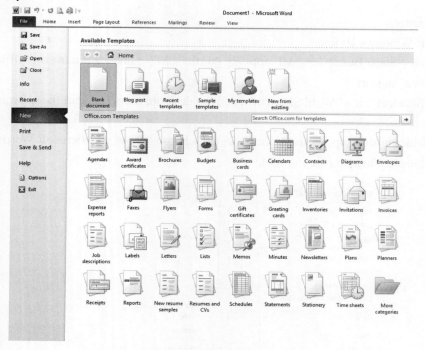

Fig 3.7 File Tab

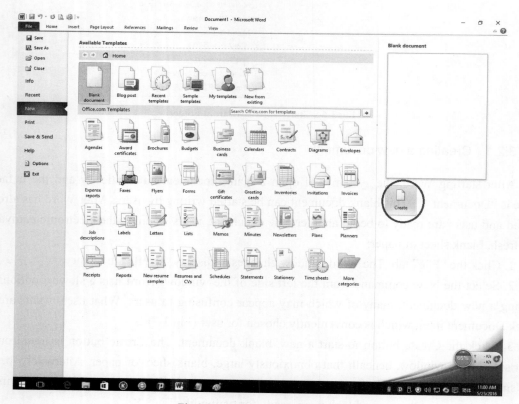

Fig 3.8 Create a new document

4. Repeat these steps as often as users need new documents.

The New Document window contains numerous options for starting something new in Word. Rather than use the Blank Document choice, a lot of folks use templates to start documents. Templates help save time by predefining document layout and formatting (and sometimes even text).

3.2.2 Using Template

All documents in Word 2010 are based on a template. When users don't specify a template, Word uses the Normal document template, NORMAL.DOTM. Word comes with a host of templates already created, as well as any templates users whip up themselves:

1. In the File tab of Word, click "New". The New menu appears. Common templates are shown at the top of the window, with online templates listed at the bottom (see Fig 3.9).

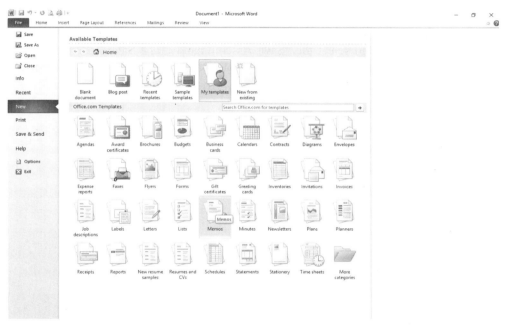

Fig 3.9 Using Template

2. If users see a template in the lists that want to use, choose it. That template becomes highlighted.

3. Click the "Create" button. Most of the time, a user will probably choose the "My Templates" item. Doing so displays the New dialog box, which lists icons representing all the templates he has created.

4. If a user chose "My Templates", choose an icon for the template he wants to use, and then click "OK" button in the New dialog box.

Instantly, Word loads up that template and starts a new document for users. The new document contains the styles and formats and perhaps even some text that's ready for users to use or edit.

If users don't see their custom template in the New dialog box, Word probably goofed up when it saved their template.

5. If users' template doesn't appear in the New dialog box, check the Documents or My Documents folder. And they need to find where Word put that template file.

6. When users locate it, "double-click" that icon, which starts a new document using the template.

7. Work with the document just like any other document in Word. A lot of the formatting and typing has been done for users.

3.2.3 Saving a document

After finish working with Word, the most important thing users can do to a Word 2010 document is save it. Create a permanent copy of what users see onscreen by saving the Word document as a file on the PC's storage system. In that way, users can use the document again, keep a copy for business reasons, publish it electronically, or just keep the thing for sentimental reasons:

1. Click the "File" tab and select the "Save As command". The Save As dialog box appears. Users need to use the Save As dialog box is when they first create a document and if they want to save a document with a new name or to a different location on disk, see Fig 3.10.

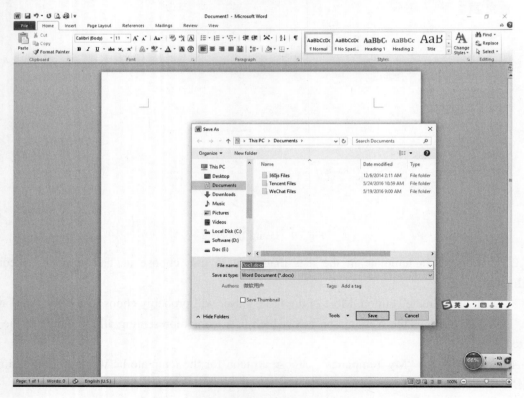

Fig 3.10 Saving a document

2. Type a name for users' document in the File Name text box. Word automatically selects the first line or first several words of the user's document as a filename and puts it in the Save dialog

box. If that's okay, users can move to Step 4. Otherwise, type a name in the File Name box.

3. (Optional) Choose a location for a file. Use the various gizmos in the Save As dialog box to choose a specific folder for a user's document.

4. Click the "Save" button. The file is now safely stored in the PC's storage system. Users clue that the file has been successfully saved is that the name users have given it (the filename) now appears on the document's title bar, near the top of the screen.

After users initially save their document by using the Save As dialog box, they can use the Save command merely to update their documents, by storing the latest modifications while they write.

3.2.4 Closing a document

Closing a Word 2010 document when users are finished working with it is simple. When users are done writing a Word document, they just need to do the electronic equivalent of putting it away. That electronic equivalent is the Close command:

1. Choose the "Close command" from the File tab menu. Alternatively, users can use the handy "Ctrl+W" keyboard shortcut. If they haven't saved their documents recently, Word prompts users to save before they close. When the document has been saved, closing it simply removes it from view like Fig 3.11.

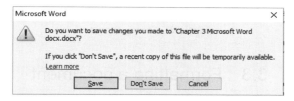

Fig 3.11　Closing window

2. If users' document needs to be saved, click the Yes button in the prompt that appears. If users haven't ever saved the document yet they see the Save As dialog box.

3. If the Save As dialog box appears, type a name for their document in the File Name text box. Word automatically selects the first line or first several words of their document as a filename and puts it in the Save dialog box. If that's okay, users can move to Step 4. Otherwise, type a name in the File Name box.

4. (Optional) Choose a location for users' file. Use the various gizmos in the Save As dialog box to choose a specific folder for their document.

5. Click the "Save" button. The file is now safely stored in the PC's storage system. Also, users can choose the format they are interested in (see Fig 3.12).

Users don't have to choose the "Close" command. They can choose the "Exit" command from the File tab menu if they are done with Word, which is almost the same thing: people are prompted to save a document if it needs saving; otherwise, the Exit Word command quits Word rather than keeps the window open.

Users can also just close the "Word program window", which closes the document. When users close the last open Word program window, users also quit Word.

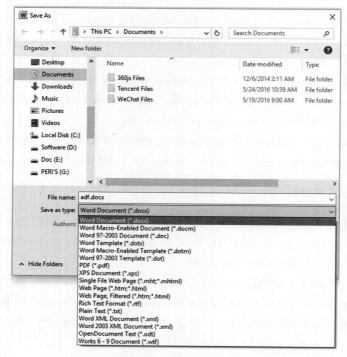

Fig 3.12　　Saving format

3.3　Formatting a document

Microsoft Word can help users change their document for a fresh look. Learn how to format text, change character and line spacing, modify paragraphs, apply borders and shading, and hide text on confidential documents. Usually, users can Format Text, Format Text for Emphasis, Change Character Spacing, Select Text with Similar Formatting, Find and Replace Formatting, Find and Replace Custom Formatting, Change Paragraph Alignment, Change Line Spacing, Display Rulers, Set Paragraph Tabs, Set Paragraph Indents, Create Bulleted and Numbered Lists, Apply Borders and Shading, Hide Text, and other several functions.

Once a user type a document and get the content how he want it, the finishing touches can sometimes be the most important. To create that interest, Microsoft Word can help users change their document for a fresh look. One of the first elements user can change is font attributes. Applying bold, underline, or italics when appropriate, can emphasize text. Users might find that having different font sizes in the document to denote various topics will also enhance the document.

Also users can change the kerning—the amount of space between each individual character, for a special effect on a title or other parts of text. They apply a dropped capital letter to introduce a body of text, add shading or border onto their document.

Word has various tools to help users format their documents. They can utilize and replace formatting effects, display rulers, change a paragraph alignment, set paragraph tabs and indents, and change line spacing.

3.3.1 Editing the text

1. Text Input: the pointer come into a blank area and a blinking "|", where is the input area (shown on Fig 3.13).

Fig 3.13　Text Input pointer

For inputting, there are many different languages Work 2010 present—English, Chinese, Deutsch, Japanese, and so on. Users can change different input language simply by using "Ctrl+Shift".

And there are some Quicklaunch and help users to operate their document.

- Change the input cursor to the beginning of documents by using "Ctrl+Home" or change to the end of documents by using "Ctrl+End".
- Change the input cursor to the beginning of the page by using "Ctrl+Alt+PageUp" or change to the end of the page by using "Ctrl+Alt+PageDown".

When a user wants to input some contents in between of words, he can click "Insert" to insert text. Or click the key again, it will insert new text and erase the old text.

2. Modifying, deleting, copying of a document

The great thing about word processors is that users can go back and edit document as much as they like, before printing it out.

For a Word user, identify the text he want to affect is the first thing he should do.

If users want to make a change to some existing text (to delete it, format it, move it), they need to identify what text is to be affected. When people select text, Word highlights the text.

To select the text, users need to identify where the selection is to start. They can do that by positioning the cursor. And then they need to identify where the selection is to end.

There are some quick ways to position the cursor:
- Click with the mouse.
- Use the arrow keys (the keys between the main area of the keyboard and the number pad). The arrow keys move one line up or down, or one character left or right.
- Ctrl + an arrow key moves one word left or right, or one paragraph up or down.
- Home moves to the beginning of the line. End moves to the End of the line.
- "Ctrl + Home" moves to the top of the document. "Ctrl+End" moves to the end of the document.

Also, there are some easy ways to select text. Position the cursor, using one of the methods shown above. Then:
- Hold down the Shift key. Click to the place for the selection to end.
- Hold down the Shift key. Use any of the methods listed above to move the cursor to create the selection.
- Double click to select one word.
- Triple click to select one paragraph.
- "Ctrl + A" to select the whole document.

To add text into a document, use the keyboard or the mouse to position the cursor where users want to add text, then type. If any text is selected, the typing will over-write the selected text.

To delete text from a document, click or use the keyboard to position the cursor. Press the "Delete" key to delete the character in front of the cursor. Press the "Backspace" key to delete the character behind the cursor. To delete a chunk of text, "select" it and press the "Delete" key.

If users want to copy text from one place to another, they can select the text they want to move. And then choose "Edit→Copy". People use the mouse or the keyboard to position the cursor where they want the text to appear and choose "Edit→Paste".

To move text from one place to another they should select the text they want to move. And then choose "Edit→Cut". Use the mouse or the keyboard to position the cursor where users want the text to appear. At last, choose "Edit→Paste" to finish moving.

To move one or more paragraphs at a time, up or down in the document, select the paragraph(s). Use "Alt+Shift+Up Arrow" to move the text up. Use "Alt+Shift+Down" Arrow to move the text down. This works for rows in tables, too.

To change the formatting of a few words, or to change the formatting of a paragraph separately from its style: first of all, select the "text" and then choose "Format→Font", "Format→Paragraph", or "Format→Borders" and Shading and make selections.

3. Find and Replace

People can quite easily in Word 2010 change every instance of one word in another word or phrase by using the Find and Replace command. How that makes the document read, of course, is anyone's guess. People may use the Find and Replace command:

On the Home tab, click the "Replace command" button, found nestled in the Editing group on the far right side. The Find and Replace dialog box appears. This place should be familiar if people have often used the Find command. After all, finding stuff is the first part of using Find and Replace (see Fig 3.14).

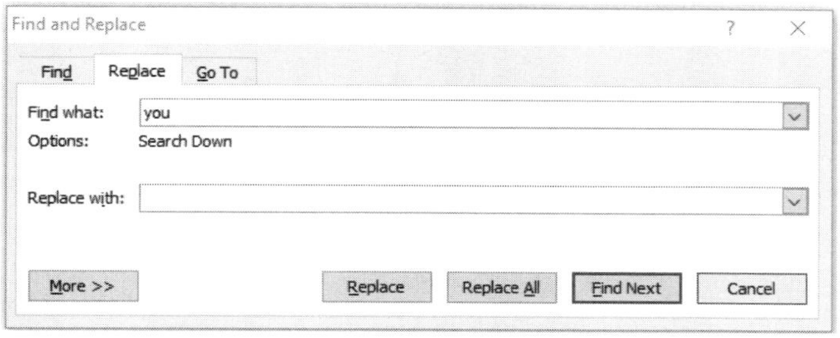

Fig 3.14 Find and Replace dialog

In the Find What box, type the text users want to find. They want to replace this text with something else. So, if they're finding misery and replacing it with company, they type "misery".

Press the "Tab" key when people complete typing. The insertion pointer jumps to the Replace With box.

In the Replace With box, type the text users want to use to replace the original text. If people don't type anything in the Replace With box, Word replaces the text with nothing!

If people click the "Find Next" button, at this point, the Replace command works just like the Find command: Word scours users' documents for the text and they typed in the Find What dialog box.

When that text is found, click the "Replace" button. Word replaces the found text, highlighted onscreen, with the text typed in the Replace With box, then it immediately searches for the next instance of the text.

Repeat click "Find Next" until the entire document has been searched. At last, users can click the "Close" button.

3. Spelling&Grammar check

Microsoft Word provides a decent Spelling and Grammar Checker which enables users to search for and correct all spelling and grammar mistakes in their document. Word is intelligent enough to identify misspelled or misused, as well as grammar errors and underlines them as follows (more details Shown in Fig 3.15).

- A red underline beneath spelling errors.
- A green underline beneath grammar errors.
- A blue line under correctly spelled but misused words.

And if users' Word cannot see those colorful underline, perhaps they need to check whether their Word turn on the Spelling&Grammer check function.

Here is the simple procedure to find out wrong spelling mistakes and fixing them in Fig 3.16.

Step (1): Click the Review tab and then click Spelling & Grammar button.

Step (2): A Spelling and Grammar dialog box will appear and will display wrong spellings or grammar and correct suggestions as shown below Fig 3.17.

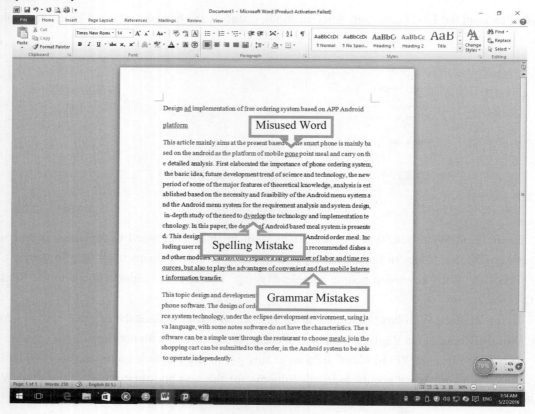

Fig 3.15 Mistakes

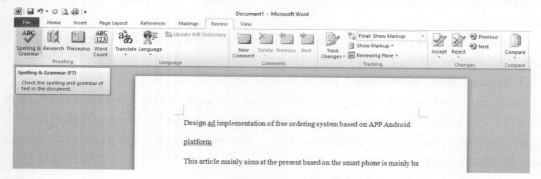

Fig 3.16 Spelling&Grammar check

- Ignore: If users are willing to ignore a word then click this button and word ignores the word throughout the document.
- Ignore All: Like Ignore, but ignores all occurrences of the same misspelling, not just this one.
- Add to Dictionary: Choose "Add to Dictionary" to add the word to the Word spelling dictionary.
- Change: This will change the wrong word using the suggested correct word.
- Change All: Like Change, but change all occurrences of the same misspelling, not just this one.

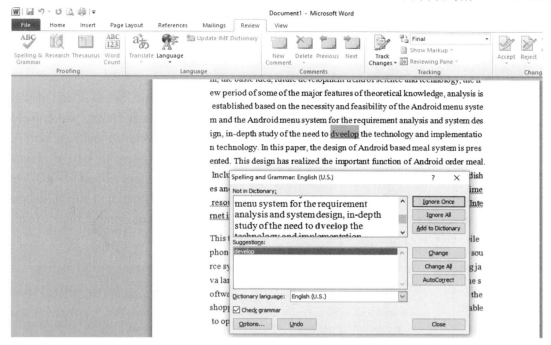

Fig 3.17　Spelling and Grammar dialog

- AutoCorrect: If users select a suggestion, Word creates an AutoCorrect entry that automatically corrects this spelling error from now on.

Following are the different options in case users have grammatical mistake:

- Next Sentence: users can click Next Sentence to direct the grammar checker to skip ahead to the next sentence.
- Explain: The grammar checker displays a description of the rule that caused the sentence to be flagged as a possible error.
- Options: This will open the Word Options dialog box to allow users to change the behavior of the grammar checker or spelling options.
- Undo: This will undo the last grammar changed.

Step (3): Select one of the given suggestions users want to use and click "Change" option to fix the spelling or grammar mistake and repeat the step to fix all the spelling or grammar mistake.

Step (4): Word displays a dialog box when it finishes checking for spelling and grammar mistakes, finally click "OK", see Fig 3.18.

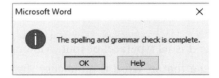

Fig 3.18　Finish checking dialog

Not every time users use Check Spelling and Grammar by those steps. Sometimes they just use "right click", If users will click a right mouse button over a misspelled word then it would show correct suggestions and above mentioned options to fix the spelling or grammar mistake.

3.3.2 Document Setting

In Microsoft Word, to change the format of documents is very simple. Select it, and change it. For setting documents, it distributes character, paragraph, and document's format, which includes changed fonts' size, color, format, space changing, as well as paragraph's editing and page setting.

1. Character formatting

The most basic element users can format in a Word 2010 document is text — the letters, numbers, and characters. People can format Word document's text to be bold, underlined, italicized, little, or big or in different fonts or colors. Word provides a magnificent amount of control over the appearance of the text, see Fig 3.19.

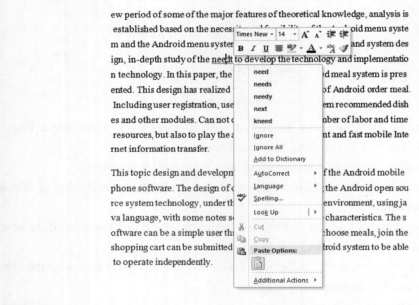

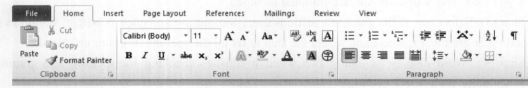

Fig 3.19　Character formatting

- Changing the font

The most basic attribute of text is its typeface, or font. Although deciding on a proper font may be agonizing, the task of selecting a font in Word is quite easy: firstly, on the Home tab, in the Font group, click the down arrow to display the Font Face list.

A menu of font options appears. The top part of the menu shows fonts associated with the document theme. The next section contains fonts users have chosen recently, which is handy for reusing fonts. The rest of the list, which can be quite long, shows all fonts in Windows that are available to Word.

- Scroll to the font users like

The fonts in the All Fonts part of the list are displayed in alphabetical order, as well as in context (how they appear when printed).

- Click to select a font

Users can also use the Font menu to preview the look of fonts. Scroll through the list to see which fonts are available and how they may look. As users move the mouse over a font, any selected text in the document is visually updated to show how that text would look in that font. (No changes are made until people select the new font.)

- Applying character formats

The Font group lists some of the most common character formats. They're applied in addition to the font. In fact, they enhance the font like Fig 3.20.

Fig 3.20　Character formatting Ribbon

- Bold: Press "Ctrl+B" or click the "Bold" command button.
- Italic: Press "Ctrl+I" or click the "Italic" command button.
- Underline: Press "Ctrl+U" or click the "Underline" command button. People can click the

"down arrow" next to the Underline command button to choose from a variety of underline styles or set an underline color like Fig 3.21.

Bold
Italic
<u>Underline</u>
~~Strike through~~

Fig 3.21 Selecting a font

To underline just words, and not the spaces between words, press Ctrl+Shift+W. Word underline looks like this.

- Strike through: Click the "Strikethrough" command button. (There's no keyboard shortcut for this one.)
- Subscript: Press Ctrl+= (equal sign) or click the "Subscript" command button. Subscript text appears below the baseline, such as the 2 in H_2O.
- Superscript: Press Ctrl+Shift+= (equal sign) or click the "Superscript" command button. Superscript text appears above the line, such as the 10 in 2^{10} like Fig 3.22.

Fig 3.22 Subscript & Superscript

There are Font box (Fig 3.23) can provide more format for text control.

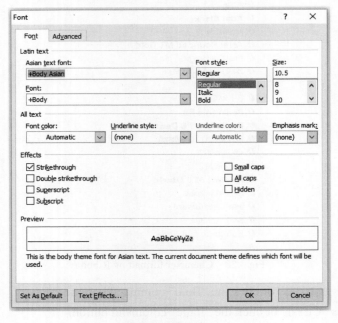

Fig 3.23 Font box

- All caps: Press "Ctrl+Shift+A". This is a text format, not applied by pressing the "Shift" or "Caps Lock key".
- Double-underline: Press "Ctrl+Shift+D". This text is double-underlined.
- Hidden text: Press "Ctrl+Shift+H". To show hidden text, click the "Show/Hide" command button (in the Paragraph Group on the Write tab). The hidden text shows up in the document with a dotted underline.
- Small caps: Press "Ctrl+Shift+K". Small caps is ideal for headings.

To turn off a text attribute, use the command again. For example, press "Ctrl+I" to type in italic. Then press "Ctrl+I" again to return to normal text.

2. Paragraph formatting

Paragraph is a unit of text or other content that starts at the beginning of a document, immediately after ad return, a page break, or a section break, or at the beginning of a table cell, header, footer, or list of footnotes and ends with a hard return (carriage return) or at the end of a table cell. Word documents generally contain paragraphs with different formatting. Even a very simple document with a centered heading and a justified body contains paragraphs with two different types of formatting.

The paragraph formatting is control the spacing, alignment, bullet and other functions.

Many options are available directly in the Paragraph group on the Home tab of the Ribbon, in the Paragraph group on the Page Layout tab, and on the contextual toolbar and menu that appear when people right-click within text (see Fig 3.24).

Fig 3.24 Paragraph formatting Ribbon

- Alignment

Alignment or justification refers to the way in which the lines of a paragraph are aligned. There are four types of alignment, and the type of alignment of the paragraph where user's cursor is located is indicated by the highlighted button in the Paragraph group on the Home tab.

> With left alignment (≡) (the default), the left-hand ends of all the lines in the paragraph are aligned along the left-hand margin of the text area.
> With center alignment (≡), the mid-points (centers) of all the lines in the paragraph are aligned along the same imaginary vertical line at the center of the text area between the margins.
> With right alignment (≡), the right-hand ends of all the lines in the paragraph are aligned along the right-hand margin of the text area.
> With justified alignment or full justification (≡), all the lines in the paragraph, except the last line, are extended so that the left-hand end of each line is aligned along the left-hand margin of the text area, the right-hand end of each line is aligned

along the right-hand margin of the text area, and the lines are all of the same length. This is achieved by inserting additional space between words.

- Line Spacing

Line spacing refers to the vertical distance between the lines within a paragraph and determines the location of each line relative to the line above it. Line spacing can be specified by name (single, 1.5 lines, double), by a number that indicates a multiple of single spacing (for example, 2.0 is equivalent to double spacing), and by an exact distance in points, where a point (pt) is equal to 1/72 of an inch. People can quickly view and change the line spacing to several common standard values by clicking the Line Spacing button ( ) in the Paragraph group on the Home tab. More line spacing options become available when users click "Line Spacing" options to open the Paragraph dialog box (see below Fig 3.25).

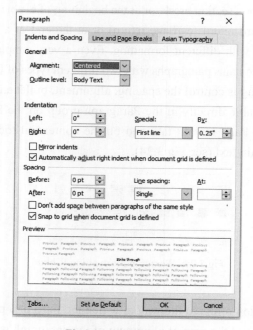

Fig 3.25　Paragraph box

- Indents

The indent before text refers to the width of the additional empty space that is inserted between the margin and the text on the left-hand side of a paragraph of left-to-right text, and the indent after text refers to the width of the additional empty space that is inserted between the text and the margin on the right-hand side of a paragraph of left-to-right text. People can quickly increase the indent before text to the next tab stop by clicking the "Increase Indent" button ( ) in the Paragraph group on the Home tab, and users can quickly decrease the indent before text to the preceding tab stop by clicking the "Decrease Indent" button ( ) in the Paragraph group on the Home tab.

- Paragraph Spacing

The spacing between paragraphs is determined by the spacing before it and the spacing after it that is set for each paragraph. People can modify the spacing before a paragraph and the spacing after it by changing the values in the applicable boxes in the Paragraph group on the Page Layout tab.

Note. When the first of two consecutive paragraphs has non-zero spacing after it and the second paragraph has non-zero spacing before it, only the larger of the two spaces will be inserted between the paragraphs (see Fig 3.25).

- Spacing

The spacing between two paragraphs is determined by the spacing before one paragraph and the spacing after the preceding paragraph, which are displayed and can be modified in the Before and After boxes. When the first of two consecutive paragraphs has non-zero spacing after it and the second paragraph has non-zero spacing before it, only the larger of the two spaces will be inserted between the paragraphs.

Line spacing refers to the vertical distance between the lines within a paragraph and determines the location of each line relative to the line above it. The following types of line spacing can be specified in the Line spacing box.

- Single
- 1.5 lines
- Double
- At least. When this option is selected, an exact distance in points, where a point (pt) is equal to 1/72 of an inch, is specified in the At box.
- Exact. When this option is selected, an exact distance in points, where a point (pt) is equal to 1/72 of an inch, is specified in the At box.
- Multiple. When this option is selected, a number that indicates a multiple of single spacing (for example, 2.0 is equivalent to double spacing), is specified in the At box.
- Borders.

If a user wants to add borders around the paragraph where the cursor is located, click the "Borders" button (Borders) to add the current default borders (the original default or the last border style that he selected). If he wants to select a border style that differs from the current default border style, click the "small arrow" on the Borders button, and select one of the border styles displayed or click Borders and Shading to define custom borders. If users want to add borders around multiple paragraphs, select the applicable paragraphs before they click the "Borders" button or the "small arrow" on it (see Fig 3.26).

Fig 3.26　Bottom Border setting

- Shading (Colored Background)

Sometimes users need to add shading with the current default background color to the entire text area of the paragraph. They can choose the "Shading" button (Shading) in the Paragraph group on the Home tab. Also, users can select a background color other than the current default color, click the small arrow on the "Shading" button and then click one of the colors displayed or define their own custom color. If they want to apply the same shading to multiple paragraphs, select the applicable paragraphs before click the "Shading" button or the "small arrow" on it.

- Numbering and bullets

Numbering is needed, when a list contains items that are in a certain order or that need to be referenced somewhere. Users can apply numbers or letters or another type of sequential marking. Users can select the paragraphs as a block and click the "Numbering" command button from the Paragraph group on the Home tab.

When users click the button, each paragraph is numbered. And then, they can use the Numbering command button's menu to choose another sequential format, such as letters or Roman numerals, or choose a specific numbering style. Or, there are none of the predefined formats in the menu can be selected by the user, choose "Define New Number Format" to create own numbered list (see Fig 3.27).

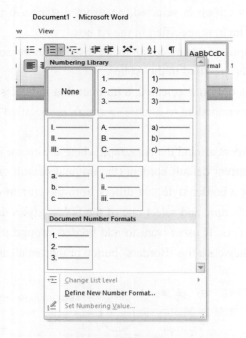

Fig 3.27　Numbering and bullets

- Making a bulleted list

In typesetting, a bullet is a graphical element, such as a ball or a dot, used to highlight items in a list. To apply bullets to text, similar with numbering users need to highlight the paragraphs and choose the "Bullets" command button, found in the Home tab's Paragraph group. In this case, text is not only formatted with bullets but is also indented and made all neat and tidy.

Also like numbering, users can choose a different bullet style by clicking the "menu" button next to the "Bullets" command. Choose their new bullet graphic from the list that appears or use the "Define New Bullet" command to dream up own bullet style (see Fig 3.28).

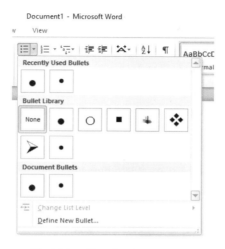

Fig 3.28 Choosing bullet's style

- Creating a multilevel numbered list

The Multilevel List button, found in the Paragraph group on the Home tab, is used to number a multileveled list, consisting of sublevels and indents (Fig 3.29).

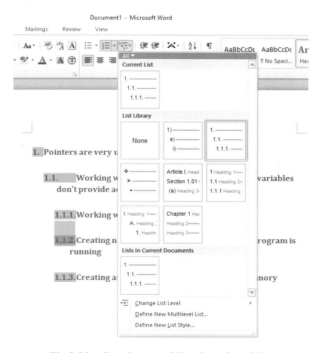

Fig 3.29 Creating a multilevel numbered list

Users can create a multilevel list from scratch, or they can apply the format to a selected block of text. The secret is to use the "Tab" key and "Shift+Tab" key combo at the start of the paragraph to shuffle the paragraphs higher and lower in the multilevel list hierarchy:

- Press the "Tab" key at the start of a paragraph to indent that paragraph to a deeper level in the multilevel list format.
- Press the "Shift+Tab" key combination at the start of a paragraph to unindent a paragraph to a higher level in the multilevel list format.
- Press the "Enter key" **twice** to end the list.

1. Styles

Styles in Word 2010 can help users to define the format style of text. Users can choose Headings, or other formats, like Fig 3.30.

When users changed as Heading 1, Heading 2, Heading 3, Normal, or Title, they can easily see the hierarchy structure with Outline Layout. Also they can check which part of text they missed, like Fig 3.31.

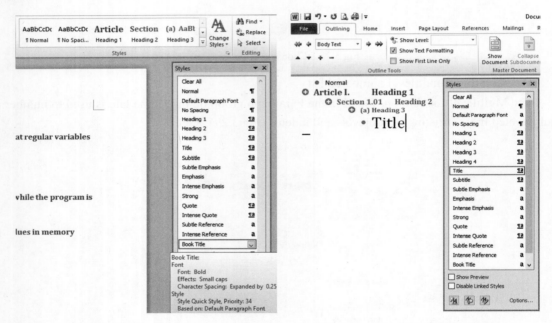

Fig 3.30　Styles　　　　　　　　　Fig 3.31　Hierarchy structure of styles

3.3.3　Pictures and Text

Users can do all kinds of things with graphics in Word 2010. Look for all the different types of graphics on the Insert tab. Images are placed inline with text, which means that they appear wherever the insertion pointer is blinking.

To choose a picture into a Word document, users can use an image file from their computer's mass storage system (Fig 3.32).

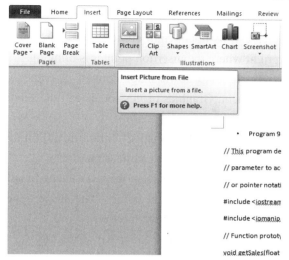

Fig 3.32   Insert Picture

From the Insert tab's Illustrations group, click the Picture button.

The Insert Picture dialog box appears. And then people use the dialog box controls to browse for the image they want. Click to select the image, and click the "Insert" button. In this point, the image is slapped down into the document.

After the user inserts a picture, the Picture Tools Format tab appears on the Ribbon. Clip art is a collection of images, both line art and pictures, that users re free to use in documents. First of all, on the Insert tab, in the Illustrations group, click the "Clip Art" button. And then the Clip Art task pane appears. Secondly, in the Search For box, type a description of what users' want. Finally, Click the "Go" button (Fig 3.33).

Fig 3.33   Insert Clip art

Users can click the image of their want. The image is plopped into document. Close the Clip Art task pane by clicking the X in its upper-right corner. Word sticks the clip art graphic into text, just like it's a big character, right where the insertion pointer is blinking.

Moreover, in Word, users can slap down shape content. It comes with a library of common shapes ready to insert into document. Graphics professionals call the shapes line art. People can call them forth into document:

1. Choose a predefined shape from the "Shapes" button menu, found in the Illustrations group on the "Insert tab". The mouse pointer changes to a plus sign.

2. Drag the mouse in the document where users want the shape to appear. Drag down, from the upper-left corner of the shape to the lower-right corner. The shape appears at the location where users draw it, as a size determined by how drag the mouse.

There is full function combination called SmartArt. Clicking the "SmartArt" button in the Illustrations group on the Insert tab and a SmartArt Graphic dialog box will pump out. People can use that dialog box to quickly arrange a layout of graphics in their document. After picking a layout, they type captions or choose images (or both). They can use the "Change Colors" button to apply some life to the otherwise dreary SmartArt. The command button is in the Quick Styles group on the SmartArt Tools Design tab (Fig 3.34 & 3.35).

1. Choose a predefined shape from the shapes button menu, found in the Illustrations group on the Insert tab. The mouse pointer changes to a plus sign.

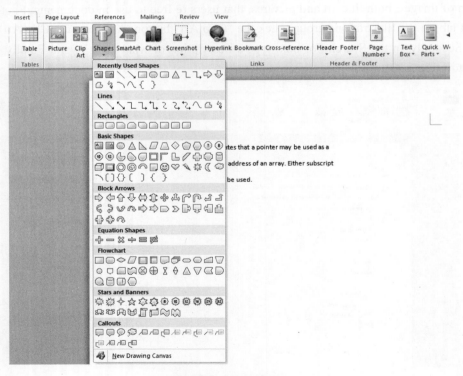

Fig 3.34   Insert Shapes

Chapter 3 Microsoft Word 2010
文字处理软件——Word 2010

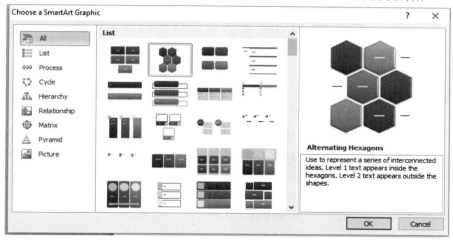

Fig 3.35 Insert SmartArts

2. Drag the mouse in the document where users want the shape to appear. Drag down, from the upper-left corner of the shape to the lower-right corner. The shape appears at the location where users draw it, as a size determined by how drag the mouse.

Perhaps the most overused graphic stuck into any Word document is WordArt.

1. On the Insert tab, in the Text group, click the "WordArt" button to display the WordArt menu.

2. Choose a style from the gallery for WordArt.

3. Type the (short and sweet) text users want WordArt-ified.

People may think text appears as an image in their documents, but it's just a graphic image, like any other graphic documents (Fig 3.36 & Fig 3.37).

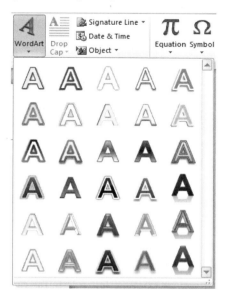

Fig 3.36 Insert WordArts

Welcome to Computer's World

Fig 3.37   An example of WordArts

Sometimes people need take a screenshot in Word. When the image a user need is on the computer screen, either in another program window or the other program window itself, user can use Word's Screenshot command to capture that image and stick it into user's document (Fig 3.38):

1. Set up the program window that user wants to appear in user's Word document. Switch to the program and position everything for picture-taking.

2. Switch back to Word.

3. Click the "Screenshot" button, found in the Illustrations group on the Insert tab.

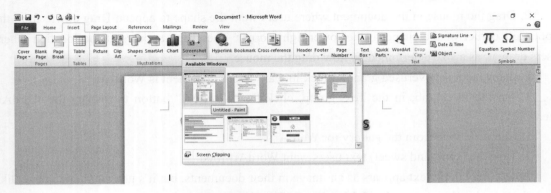

Fig 3.38   Screenshot's example

A stubby little menu appears. It lists any other program windows that are open and not minimized.

3.3.4   Table

A table is an element people insert into their document, so Word 2010's Table commands are found on the Ribbon's Insert tab, in the named Tables group. Word comes with an assortment of predefined, formatted tables. Plopping one down in users' document is as easy as using the Quick Tables submenu, chosen from the Table menu on the Insert tab (Fig 3.39).

First of all, click the Table button on the Insert tab and choose Draw Table from the menu that appears. The insertion pointer changes to a pencil (the "pencil pointer" see Fig 3.40).

Secondly, users can click in the document and drag to "draw" the table's outline. Start in the upper-left corner of where users envision their table and drag to the lower right corner, which tells Word where to insert the table. Users see an outline of the table while they drag down and to the right. If people need to draw a row, drag the pencil pointer from the left side to the right side of the table. As long as the mouse pointer looks like a pencil, users can use it to draw the rows and columns in a table.

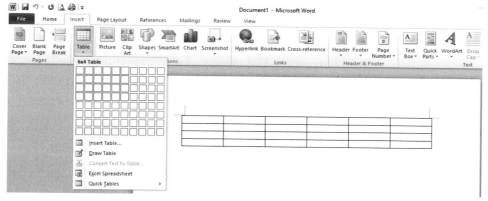

Fig 3.39   Insert Table

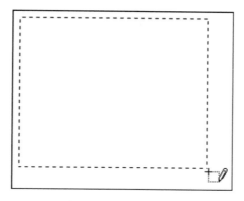

Fig 3.40   Drawing table

Also, to draw a column, drag the pencil pointer from the top to the bottom of the table. Users can split columns or rows into more cells by simply dragging the pencil pointer inside a cell and not across the entire table. Click the "Draw Table" button or press the "Esc" key when users are done creating the table's columns and rows. In this point, the mouse pointer returns to normal. Users can begin putting text in the table.

Last but not least, if users need to draw more lines in a table, click the "Draw Table" button in the Design tab's Draw Borders group. The table people modify need not be created by the Draw Table command and any table can be modified by using that tool.

Sometimes, users may add a row, adjusting the width of a table element — they can use Word's Table Tools tabs after the table has been created. The Table Tools tabs show up only when a table is being edited or selected. And the best time to format and mess with a table is after people finish putting text into the table.

Manipulating a Word table with the mouse, for quick-and-dirty table manipulation, users can use the mouse. And Clicking-and-dragging the mouse on a vertical line in the table's grid allows users to adjust the line left or right, and resize the surrounding cells. They can also adjust cell width by using the Ruler, by pointing the mouse at the "Move Table Column" button that appears above each table cell gridline. Also, clicking-and-dragging the mouse at a horizontal line allows users to adjust the line up or down, and change the row height of surrounding cells.

It's the Table Tools Layout tab that harbors many of the command buttons and items that let users to manipulate and adjust a table. People can start table changing by placing the insertion pointer somewhere within the table itself, which makes the Table Tools Layout tab (among others) appear (like Fig 3.41):

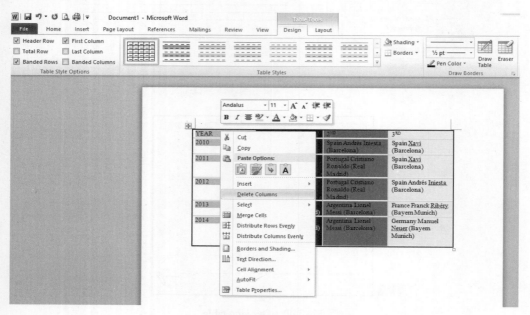

Fig 3.41   Table Layout

Deleting cells, columns, or rows: Position the insertion pointer in the part of the table users want to remove, then right click mouse to choose the table element to remove. When users choose the Delete Cells command, they can see a dialog box asking what to do with the other cells in the row or column: move them up or to the left.

If people need to insert columns or rows: Four commands in the Rows & Columns group make this task possible: Insert Above, Insert Below, Insert Left, and Insert Right. The row or column that's added is relative to where the insertion pointer is within the table.

For adjusting row and column size, there are some tools in the Cell Size group help users fine-tune the table's row height or column width. The Distribute Rows and Distribute Columns command buttons, found in the Cell Size group, help clean up uneven column or row spacing in a table.

Text within a cell can be aligned just like a paragraph: left, center, or right. Additionally, the text can be aligned vertically: top, middle, or bottom.

"The Text Direction" button in the Alignment group changes the way text reads in a cell or group of selected cells. Normally, text is oriented from left to right. By clicking the "Text Direction" button once, users change the text direction to top-to-bottom. Click the button again and direction is changed to bottom-to-top. Clicking a third time restores the text to its normal direction.

Perhaps user need to design a Word table and the Table Tools Design tab is used to help users quickly (or slowly) format their table. The tab shows up whenever the insertion pointer lies somewhere in a table's realm (like Fig 3.42):

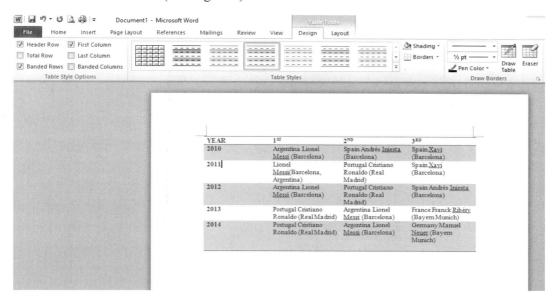

Fig 3.42 Table Design

The fig showing above is using Quick Styles. The Table Styles group can quickly apply formatting to any table. Choose a style or click the menu button to see a whole smattering of styles. The Quick Styles don't work when user have a table in a document created by or saved in an older version of Word.

There are some methods to modify the format of users table:
- Setting table line styles: The lines users see in a table's grid are the same borders they can apply to text with the Border command button, which determines where the lines go; items on the left side of the Draw Borders group set the border line style, width, and color.
- Removing a table's lines: Occasionally, users may want a table without any lines. Select the table and choose No Border from the Borders menu. Having no lines (borders) in a table makes working with the table more difficult. The solution is to show the table gridlines, which aren't printed. To do that, select the table and choose the Show Gridlines command from the Border menu.
- Merging cells: users can combine two or more cells in a table by simply erasing the line that separates them. To do so, click the Eraser command button found in the Draw Borders group. The mouse pointer changes to a bar of soap, but it's supposed to be an eraser. Click a line and it's gone. Click the Eraser button again when they're done merging (see Fig 3.43).
- Splitting cells: To turn one cell into two, users simply draw a line, horizontally or vertically, through the cell. Click the "Draw Table" command button in the Draw Borders group, and then

draw new lines in the table. Click the "Draw Table" button again to turn off this feature.

Fig 3.43   Merging cells

When users change their mind and need to delete a Word table, they have to click the mouse inside the table and then choose Delete Table from the Rows & Columns group on the Layout tab. The table is blown to smithereens.

Microsoft Word allows users to place a border on any or all of the four sides of a table very similar to text, paragraphs, and pages. People can also add many type of shading to table rows and columns. How to add any of the borders (left, right, top or bottom) around a table and how to add different shadows to various rows and columns of the table, users have to consider in following methods.

- Add Borders To Table:
  - Select the table to which users want to add border. To select a table, click over the table anywhere which will make Cross Icon visible at the top-left corner of the table. Just click this cross icon to select the table.
  - Click the "Border Button" to display a list of options to put a border around the selected table. Users can select any of the option available by simply clicking over it (see Fig 3.44).
  - Try to add and remove different borders like left, right top or bottom by selecting different options from the border options.
  - User can apply border to any of the selected row or column.
  - To delete the existing border, simply select "No Border" option from the border options.

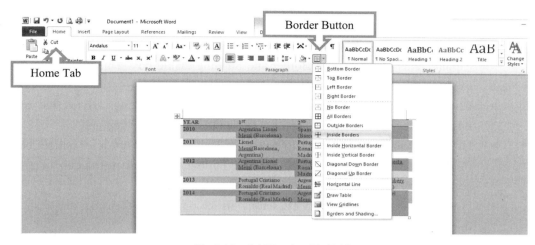

Fig 3.44   Add Borders To Table

- Using Border Options (more detionls on Fig 3.45):

Fig 3.45   Using Border Options

Users can add borders of their choice to word table by following the simple steps given below.
- ➢ Click the Border Button to display a list of options to put a border. Select Border and Shading option available at the bottom of list of the options as shown in above screen capture. This will display a Border and Shading dialog box. This dialog box can be used to set borders and shading around a selected table.
- ➢ Click Border tab which will display a list of border settings, styles and options whether this border should be applied to the table or text or paragraph.
- ➢ Users can use Preview section to disable or enable left, right , top or bottom borders of the selected table or row or column. Follow the given instruction in preview section itself to design the border people like.
- ➢ Users can customize border by setting its color, width by using different width thickness available under style section.

- Add Shades To Table:

People occasionally need use shading to emphasis one cell, column, or row, so they can add shades on a selected table.
- ➢ Select a row or column where a user wants to apply shade of the choice (Fig 3.46).

YEAR	1ST	2ND	3RD
2010	Argentina Lionel Messi (Barcelona)	Spain Andres Iniesta (Barcelona)	Spain Xayi (Barcelona)
2011	Lionel Messi (Barcelona, Argentina)	Portugal Cristiano Ronaldo (Real Madrid)	Spain Xayi (Barcelona)
2012	Argentina Lionel Messi (Barcelona)	Portugal Cristiano Ronaldo (Real Madrid)	Spain Andres Iniest (Barcelona)
2013	Portugal Cristiano Ronaldo (Real Madrid)	Argentina Lionel Messi (Barcelona)	France Franck Ribery (Bayem Munich)
2014	Portugal Cristiano Ronaldo (Real Madrid)	Argentina Lionel Messi (Barcelona)	Gerrnany Manuel Neuer (Bayem Munich)

Fig 3.46 An example of table shading

- ➢ Click the Border Button to display a list of options to put a border. Select Border and Shading option available at the bottom of list of the options. This will display a Border and Shading dialog box. This dialog box can be used to set borders and shading around selected row(s) or column(s) (see Fig 3.47).
- ➢ Click "Shading" tab which will display options to select fill, color and style and whether this border should be applied to cell or table or selected text.
- ➢ People can use Preview section to have an idea about the expected result. Once they are done, click "OK" button to apply the result.

Chapter 3　Microsoft Word 2010
文字处理软件——Word 2010

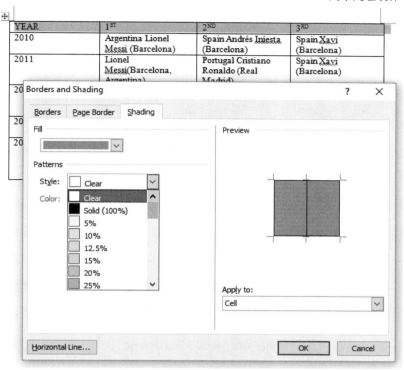

Fig 3.47　Set borders and shading

3.4　Finalizing a Document

When users finish typing their text and all tables, they need to finalize their document to make those works look formatted and professionalized. People still have to use some function to achieve this.

3.4.1　Page Design

1. Header and footers

Users can add headers and footers to Word 2010 documents. A header is text that appears at the top of every page. It's contained in a special, roped-off part of the page where people can place special text. A footer is text that appears at the bottom of every page. Like the header, it has its own, special area and contains special text. People can use a header, footer, or both (Fig 3.48):
- Click the "Insert" tab and, from the Header & Footer group, click the "Header" button. A list of preformatted headers is displayed.
- Select the format from the list. The header is added to the user document, saved as part of the page format. If Word is in Draft view, it immediately switches to Print Layout view so that the user can edit the header. (Headers and footers don't appear in Draft view.) Also, the Header & Footer Tools Design tab appears at the top of the Word window.

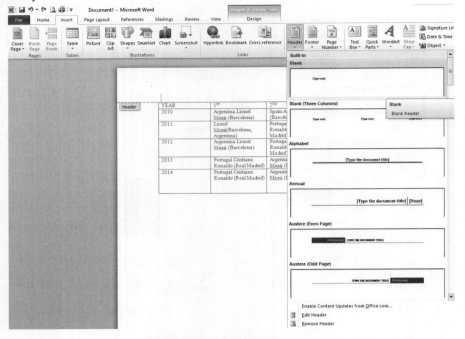

Fig 3.48　Header and footers

- Click any bracketed text and type the required replacement text. For example, replace [Type text] with the title of a document (see Fig 3.49).

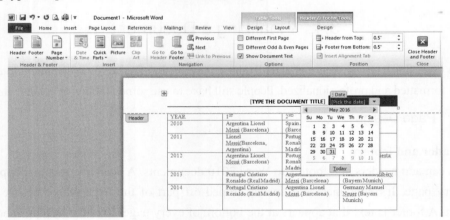

Fig 3.49　Change the title of Header

- When users finish editing, click the "Close Header and Footer" command button in the Close group on the far right side of the Ribbon.

Therefore, users can edit the header using what Word created as a starting point, or users can just quickly whip up their own header.

- Click the "Insert" tab and, in the Header & Footer group, choose Header→Edit Header. Or, in Print Layout view, people can quickly edit any header or footer by double-clicking its ghostly gray image (see Fig 3.50).

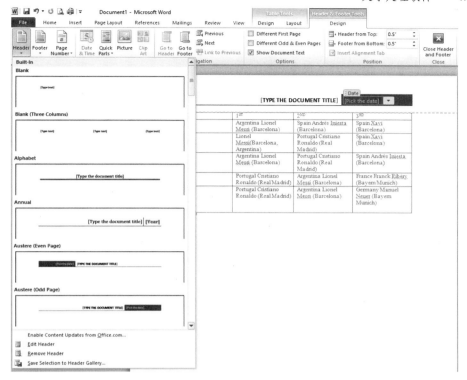

Fig 3.50　Edit Header

When people edit the header, Word provides a special mode of operation. The header appears on the screen as part of the page in Print Layout view. Plus, the Header & Footer Tools Design tab appears with groups customized for creating and editing headers.

- If necessary, click the "Go to Header" command button. Word switches to the header for editing.
- Edit or modify the header. Items in the header are edited just like anything else in Word. People can add or edit text, and format that text by using any of Word's text and paragraph-formatting commands, including tabs.
- Use the command buttons on the Design tab's Insert group for special items. For example, people can insert into a header a Page Number, Date & Time, Graphics, and Fields.
- If people have a footer to modify, click the "Go to Footer" command button, and then edit or modify the footer. This process is identical to how people modify a header.
- Click the "Close Header and Footer" command button when people are done.

2. Page Break and Section Break

Page and section breaks probably cause the most confusion and trouble for the untrained user. Documents end up with unwanted breaks that play havoc with page numbering, formats, and printing. Users don't always realize that they're the problem—they inserted the breaks, whether intentionally or not (see Fig 3.51).

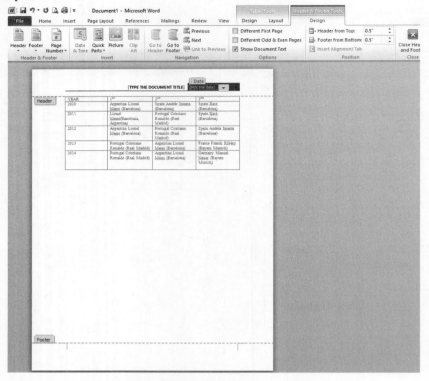

Fig 3.51　Insert Page Number, Date & Time, Graphics, and Fields

People can create a new page at any time by pressing "Ctrl+Enter". Or, click the "Page Break" option in the Pages group on the Insert tab. Unfortunately manual page breaks (also known as hard page breaks) cause trouble because they don't flow with the document's structure. As people add and delete elements, users might find manual page breaks no longer appropriate. Fortunately, they're easy to delete. Position the cursor at the beginning of the next page and press "Delete". Or, click the "Show/Hide" option in the Paragraph group on the Home tab to display the page break element, highlight it, and press "Delete". Manual breaks are probably the easiest break problem to find and resolve. Manual page breaks might be easy to insert, but they're seldom the best way to break. Sometimes the break really belongs to the text. That happens when users want a break to occur before or after a specific paragraph of text. Consequently, people could end up with an unexpected page break that's all but impossible to get rid of, unless people know its cause. To access these options, click the "Paragraph group's dialog" launcher and then click "the Line and Page Breaks" tab.

Word enables the Window/Orphan Control by default. This option prevents a single line from appearing at the top or bottom of a page. The remaining options, which people will apply as needed, follow:

- Keep With Next: This option glues the current paragraph to the following paragraph. If Word has to move both paragraphs to the next page to do so, it will. (This option is part of the built-in heading styles.) Users can use this option to keep introductory text and

headings with the text that immediately follows.
- Keep Lines Together: This is the simplest option to understand; it keeps all the lines of the current paragraph together on the same page.
- Page Break Before: This little-used option forces Word to begin the current paragraph at the beginning of a new page, like Fig 3.52.

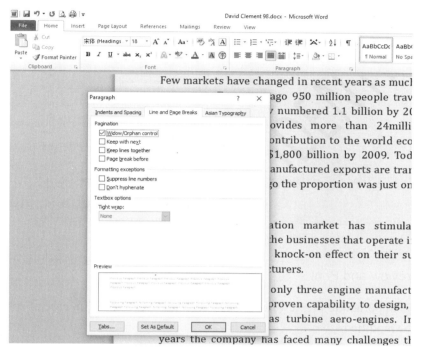

Fig 3.52　Page Breaks

　　Section breaks can be more troublesome than page breaks, because many users don't understand the nature of sections. A section lets users control formatting as needs change. For instance, people might want to print part of an entire page in landscape in the middle of a document that's using portrait orientation. To do so, users insert a new section for the landscape components and format that section as landscape. The sections before and after would remain in portrait. To access section breaks, click the "Page Layout" tab.

　　Then, choose the appropriate option from the Breaks option in the Page Setup group:
- Next Page: Starts the new section on the next page.
- Continuous: Starts the new section on the same page.
- Even Page: Starts the new section on the next even-numbered page.
- Odd Page: Starts the new section on the next odd-numbered page.

　　Issues sometimes arise when users apply Next Page rather than Continuous because Word inserts a page break, and that might not be what the user wanted.

　　Also, a column break, accessible via the Breaks option (see Fig 3.53) pushes columnar text to the next column. It breaks a column, not a page, but sometimes, it does create an automatic page break. When users insert a column break in the last column on a page, the break also acts as a page break. It makes sense as Word can't push the text into a column that doesn't exist - the only option is the next page. If people don't want the page break, they'll have to delete the column break.

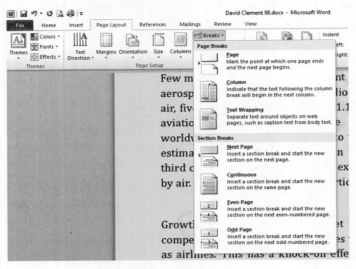

Fig 3.53   Choosing Breaks

3.4.2 Preview and Print

It is always important to preview documents before print them out, in case of some mistakes or points need to be modified.

In Microsoft Office 2010 programs, people now preview and print their Office files in one location — on the Print tab in the Microsoft Office Backstage view.

3.4.3 PDF Conversion

PDF is Portable Document Format and is standard for sharing documents on the web. In so many situations, people need to create and test an accessible PDF made from a Word 2010 document.

An accessible Word document needs have proper heading styles, list styles for numbered and bulleted lists, and alternative text for images. When users convert the document to PDF using the proper save options, the same heading levels, list styles, and alt text will be found in the resulting PDF file.

The easiest way is saving Word document to PDF, use either the Word Save As function or the Adobe Acrobat Create PDF plugin. People are using the Word "Save As" Function, the most common method to use is the Word "Save As" functions since no additional software is required. To convert people's Word document to PDF: Choose the Save As option, either in the File tab or the Productivity/Accessibility tab on the Word ribbon. And then in the Save As dialog, choose the PDF option in the Save As type field (see Fig 3.54 & 3.55).

Chapter 3　Microsoft Word 2010
文字处理软件——Word 2010

Fig 3.54　Print Preview

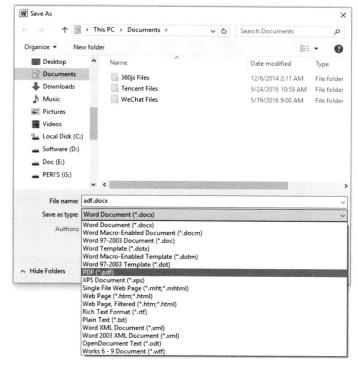

Fig 3.55　PDF Conversion

Chapter 4　Microsoft Excel 2010
电子表格 Excel 2010

In the former chapter, people can use simple tables to calculate or arrange data. In this chapter, it presents a spreadsheet format for easily data input, calculation, disposal and analysis.

4.1　An introduction of MS Excel 2010

Microsoft Excel is a powerful spreadsheet that is easy to use and allows users to store, manipulate, analyze, and visualize data by using a grid of cells arranged in numbered rows and letter-named columns to accomplish arithmetic operations. It includes many powerful tools that can be used to organize and manipulate massive amount of data. It is not only arithmetic functions Excel can present, but statistical, engineering and financial needs can be included in Excel. In addition, it can display data as line graphs, histograms and charts, and with a very limited three-dimensional graphical display.in more professional method, Excel can be applied automatically poll external databases and measuring instruments using an update schedule, analyzed the results, made a Word report or PowerPoint slide show, and e-mailed these presentations on a regular basis to a list of participants. Even now, Excel can cooperate with the newest technology for data disposal—the Cloud calculation!

4.1.1　Excel 2010 Components

Microsoft Excel is an electronic spreadsheet that can be used to organize data rows and columns, to perform mathematical calculations quickly and even can also be programmed to send mails at pre-defined time.

The Microsoft Excel 2010 window appears and our screen looks similar to the one shown here.

Screen always might not look exactly like the screen shown. In Excel 2010, display of window depends on the size of the monitor and the resolution to which the monitor is set. Resolution determines how much information the computer monitor can display. A low resolution means less information fits on the screen, but the size of text and images are larger. Inversely, a high resolution means more information fits on the screen, but the size of the text and images are smaller. Also, settings in Excel 2010, Windows Vista, and Windows XP allow users to change the color and style of their windows.

- The Worksheet

Microsoft Excel 2010 consists of worksheets (Default Names are "Sheet1", "Sheet2" and "Sheet3"). Each worksheet contains columns and rows. The columns are lettered A to Z and then continuing with AA, AB, AC to AZ and then continuing with AAA, AAB and so on up to XFD (total columns are 16384); the rows are numbered 1 to 1,048,576 (Fig 4.1).

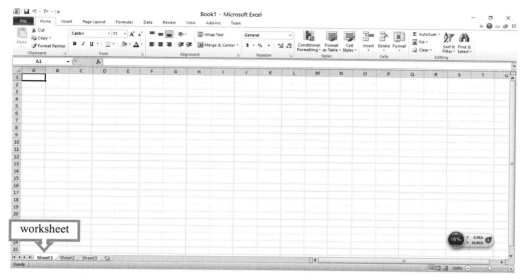

Fig 4.1   Excel Window

The combination of a column coordinate and a row coordinate defines a cell address. For example, the cell located in the upper-left corner of the worksheet is cell A1, meaning column A, row 1. Cell C7 is located under column C on row 7. People enter data into the cells on the worksheet.

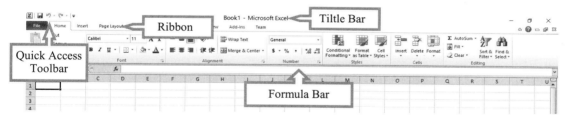

Fig 4.2   Worksheet Ribbon

- The Microsoft Office Button

Look at the upper-left corner of the Excel 2010 window, it is the Microsoft Office button. When users click this button, a menu will appear. This menu can be used to control the Excel window's appearance (Fig 4.3).

Fig 4.3   The Microsoft Office Button

- The Quick Access ToolBar

Right to the Microsoft Office button is the Quick Access toolbar. The Quick Access toolbar gives access to commands we frequently use. By default, Save (to save files), Undo (to rollback an action), and Redo (to reapply a rolled back action) appear on the Quick Access toolbar.

- The Title Bar

Next to the Quick Access toolbar is the Title bar. On the Title bar, Microsoft Excel displays the name of the workbook, which is currently in use.

- The Ribbon

User use commands to tell Microsoft Excel what to do. In Microsoft Excel 2010, people use the Ribbon to issue commands. The Ribbon is located near the top of the Excel window and below the Quick Access toolbar. At the top of the Ribbon are several tabs; clicking a tab displays several related command groups. Within each group are related command buttons. Buttons are clicked to issue commands or to access menus and dialog boxes. People may also find a dialog box launcher in the bottom-right corner of a group. When users click the dialog box launcher, a dialog box makes additional commands available.

- The Formula Bar

If the Formula bar is turned on, then in the Name box (located on left side) it displays the cell address of the cell are in. Cell entries which can be a values or formulas are displayed on the right side of the Formula bar. To turn on the Formula bar in Excel 2010 window, perform the following steps:

1. Click the "View" tab.
2. Click "Formula Bar" in the Show/Hide group. The Formula bar appears (see Fig 4.4).

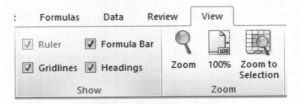

Fig 4.4   Showing Formula bar

- The Status Bar

The Status bar appears at the bottom of the Excel 2010 window and provides quick information such as the count, sum, average, minimum, and maximum value of selected numbers. Users can change what displays on the Status bar by right-clicking on the Status bar and selecting the options they want from the Customize Status Bar menu. We just need to click a menu item to select it and click it again to deselect it. A check mark appearing next to an item means the item is selected.

4.1.2 Start an Excel

Microsoft Office Excel 2010 provides several methods for starting and exiting the program. People can open Excel by using the Start menu or a desktop shortcut. When users want to exit Excel, they can use the "File tab", the "Close" button, or a keyboard shortcut (Fig 4.5).

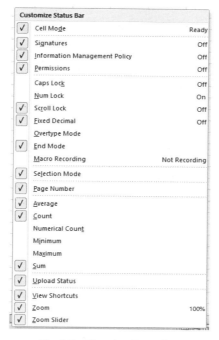

Fig 4.5   Showing Status bar

To start Excel 2010 from the Windows Start menu, choose Start→All Programs→Microsoft Office→Microsoft Excel 2010. A new, blank workbook appears, ready for users to enter data.

Sometime, if people use Excel all the time, they may want to make its program option a permanent part of the Windows Start menu. They can follow these steps to pin Excel 2010 to the Start menu:

- Click the "Start" button and then right-click Microsoft Excel 2010 on the Start menu to open its shortcut menu. If people don't see Microsoft Excel 2010 displayed on the recently used portion on the left side of the Windows Start menu, start Excel 2010 and then repeat this step like Fig 4.6.
- Click "Pin to Start" menu on the shortcut menu. After pinning Excel to the Start menu, the Microsoft Excel 2010 option always appears in the left column of the Start menu and users can then launch Excel by clicking the Start button and then clicking this option.
- Creating an Excel 2010 desktop shortcut.

People may prefer having the Excel 2010 program icon appear on the Windows desktop so that they can launch the program from there. To create an Excel 2010 desktop shortcut, follow these steps:

1. Choose Start→All Programs→Microsoft Office.

2. Right-click Microsoft Excel 2010, highlight Send To on the shortcut menu, and click Desktop (Create Shortcut) on its continuation menu.

A shortcut named Microsoft Excel 2010 appears on users' desktop. They can rename the shortcut to something shorter, such as Excel 2010.

Fig 4.6   Start Excel from Start Button

3. Right-click the Microsoft Excel 2010 icon on the desktop and then click Rename on the shortcut menu.

4. Replace the current name by typing a new shortcut name, such as Excel 2010, and then click anywhere on the desktop.

- Exiting Excel 2010

When people are ready to quit Excel, they also have several choices for shutting down the program. For example, they can choose "File→Exit", press "Alt+F4", or click the "Close" button (the X button) in the upper-right corner of the Excel 2010 program window.

If people try to exit Excel after working on a workbook and they haven't saved their latest changes, Excel displays an alert box asking whether they want to save their changes. To save their changes before exiting, people click the "Save" button. If users don't want to save changes, they can click "Don't Save".

The Backstage view has been introduced in Excel 2010 and acts as the central place for managing users' sheets. The backstage view helps in creating new sheet, saving and opening sheets, printing and sharing sheets, and so on.

People need to click the "File" tab, which located in the upper-left corner of the Excel. The opened sheets are as follows:

If user already has an opened sheet then it will display a window showing detail about the opened sheet as shown above. Backstage view shows three columns when users select most of the available options in the first column, like Fig 4.7.

Chapter 4 Microsoft Excel 2010
电子表格 Excel 2010

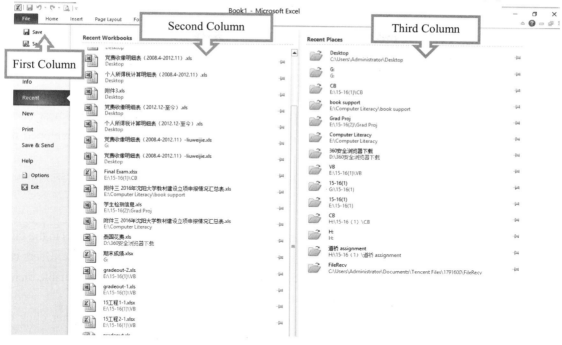

Fig 4.7 File tab

First column of the backstage view will have following options:

Table 4.1 meaning of the first column

Option	Description
Save	If an existing sheet is opened, it would be saved as is, otherwise it will display a dialogue box asking for sheet name
Save As	A dialogue box will be displayed asking for sheet name and sheet type, by default it will save in sheet 2010 format with extension .xlsx
Open	This option will be used to open an existing excel sheet
Close	This option will be used to close an opened
Info	This option will display information about the opened sheet
Recent	This option will list down all the recently opened sheets
New	This option will be used to open a new sheet
Print	This option will be used to print an opened sheet
Save & Send	This option will save an opened sheet and will display options to send the sheet using e-mail etc
Help	Users can use this option to get required help about excel 2010
Options Use	This option to set various option related to excel 2010
Exit	Use this option to close the sheet and exit

- Sheet Information

When users click Info option available in the first column, it displays the following information in the second column of the backstage view:

Compatibility Mode: If the sheet is not a native excel 2007/2010 sheet, a Convert button appears here, enabling users to easily update its format. Otherwise, this category does not appear.

Permissions: users can use this option to protect excel sheet. People can set a password so that nobody can open their sheet, or they can lock the sheet so that nobody can edit the sheet.

Prepare for Sharing: This section highlights important information people should know about the sheet before send it to others, such as a record of the edits they made as people developed the sheet.

Versions: If the sheet has been saved several times, people may be able to access previous versions of it from this section.

- Sheet Properties

When people click Info option available in the first column, it displays various properties in the third column of the backstage view. These properties include sheet size, title, tags, categories etc. Users can also edit various properties. If people try to click on the property value and if property is editable then it will display a text box where they can add their text like Title, Tags, Comments, Author.

If people finish their work in Backstage View, it is simple to exit from there. Either click on "File tab" or press "Esc" button on the keyboard to go back in excel working mode.

4.2 Creating a Worksheet

Three new, blank sheets always open when users start Microsoft Excel. But suppose that users want start another new worksheet while they are working on another worksheet or they closed already opened worksheet and want to start a new worksheet. They can choose insert a new worksheet.

4.2.1 Creating a new Worksheet

When people need a new worksheet, they can right click the Sheet Name and select "Insert" option (see Fig 4.8).

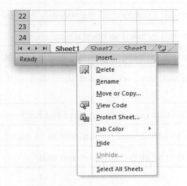

Fig 4.8 Creating a new Worksheet

And then people will see the Insert dialog with select Worksheet option as selected from the general tab. Click "OK" button (see Fig 4.9).

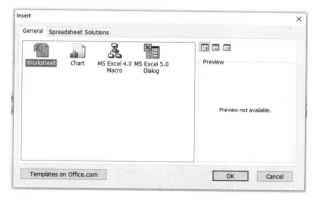

Fig 4.9　Choosing a new Worksheet

Thus, people should have their blank sheet as shown below ready to start typing their text (Fig 4.10).

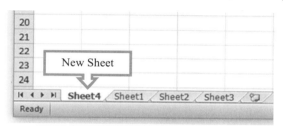

Fig 4.10　New Sheet appears

Users can use a short cut to create a blank sheet anytime. They also can use "Shift+F11" keys and a new blank sheet similar to above sheet is opened.

A worksheet is a collection of cells where users keep and manipulate the data. As showing each Excel workbook contains three worksheets above by default.

Therefore, there are some operations that need user to do.

- Select a Worksheet

When users open Excel, Excel automatically selects Sheet1 for them. The name of the worksheet appears on its sheet tab at the bottom of the document window. To select one of the other two worksheets, simply click on the sheet tab of Sheet2 or Sheet3.

- Rename a Worksheet

By default, the worksheets are named Sheet1, Sheet2 and Sheet3. To give a worksheet a more specific name, execute the following steps. First, users should right click on the sheet tab of Sheet1. And then Choose Rename. Last, people can rename a Worksheet, see Fig 4.11.

For example, Sheet4 can rename to My First Table.

- Move a Worksheet

To move a worksheet, click on the sheet tab of the worksheet users want to move and drag it into the new position.

For example, click on the sheet tab of Sheet5 and drag it before Sheet2, see Fig 4.12.

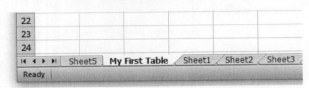

Fig 4.11   Rename a Worksheet

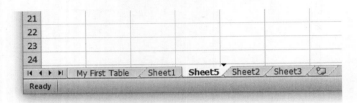

Fig 4.12   Move a Worksheet

- Delete a Worksheet

To delete a worksheet, right click on a sheet tab and choose Delete. For example, delete Sheet2 and Sheet3, see fig 4.13.

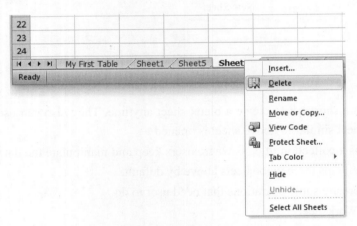

Fig 4.13   Delete a Worksheet

After click "Delete", the chosen Sheet will disappear.

4.2.2   Fill in Data

Like former Excel version, people use cell to fill in data. A cell is the rectangle formed by the intersection of a column and row. The active cell is the cell users can currently edit or modify, and it is marked with a black outline.

The easiest example of a linear series of numbers is 1, 2, 3, 4, and 5. In a linear series, the next number in the series is always obtained by adding a constant, or step value to the previous number. Said another way, each subsequent number is incremented by the same value, such as Fig 4.14.

	A	B	C	D	E	F	G
1			Student Transcript				
2	ID	Name	Ad Math	English	Computer	C	total
3	200193001	John	65	86	69	77	
4	200193002	Tersa	78	80	95	85	
5	200193003	Jack	72	53	56	60	
6	200193004	Tom	79	82	88	89	
7	200193005	Jessica	81	77	82	85	
8	200193006	Faye	88	81	89	87	
9	200193007	Alica	90	84	92	94	
10	200193008	Bob	75	70	90	85	
11	200193009	Simon	86	78	94	83	
12	200193010	Eric	51	76	49	74	
13	200193011	Mike	82	80	87	91	
14	200193012	Nicole	90	75	81	90	
15	200193013	Peter	93	82	79	85	
16	200193014	Merissa	97	91	93	89	
17	200193015	Zac	84	79	73	50	
18	200193016	Michelle	92	89	97	90	
19	200193017	Sean	79	88	88	67	
20							

Active Cell E20

Fig 4.14 Active Cell

A linear series can consist of decimals (1.5, 2.5, 3.5, 4.5), decreasing values (100, 98, 96, 94), or negative numbers (-1,-2,-3,-4). But in each case we ADD (or subtract) the constant or step value.

So, how users can autofill a column using the Fill Handle, this is an important tip for each one.

Firstly users need to open the Excel worksheet. In the first cell enter 1, and enter 2 in the cell immediately below. Highlight the two cells, and hover the cursor over the bottom right corner until they see the Fill Handle (+). After this, they press the left mouse button and drag down the column as far as users want Excel to auto fill. At the end, release the mouse button and Excel fills the column with the linear series as shown in Fig 4.15.

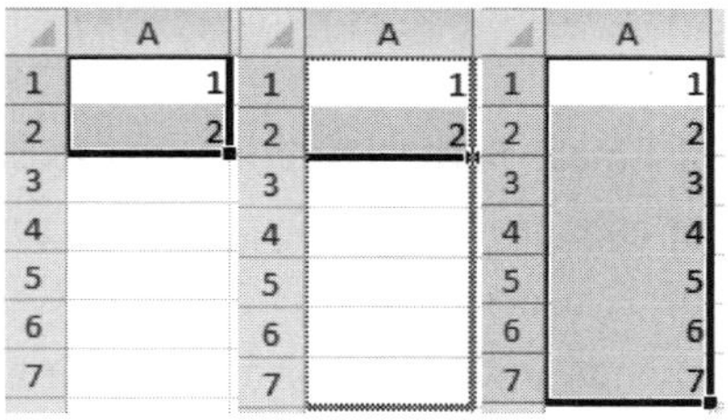

Fig 4.15 Show how to use the Excel fill handle to autofill a simple linear series in a worksheet

There also is an Alternate Option. If users want a linear series with a step value of 1, they can enter the first number only, click and drag the Fill Handle with the right mouse button, and click Fill Series on the small menu that displays. The default step value for Fill Series is one.

If users have trouble using the Fill Handle, they may accomplish the same task using the Fill button on the Editing section of the ribbon's Home tab. And then, people can base on the following steps to use the Fill button on the ribbon:

- Enter the two numbers in the worksheet.
- Highlight these numbers and the rest of the column people want autofilled.
- Click the "Fill" Button, and click Series.
- In the Series window, make sure "Linear Series" is selected with a step value of 1, and click "OK".

There are more Examples of Auto Fill of a Linear Series by using the Fill Handle to autofill numbers or other data in an Excel worksheet. Look at the sample worksheet below Fig 4.16. In each case users entered the first two numbers or other data, and dragged the Fill Handle with the left mouse button.

	A	B	C	D	E	F	G
1	1	2	5	1000	Jan	8:00 AM	Monday
2	3	6	10	2000	Feb	9:00 AM	Tuesday
3	5	10	15	3000	Mar	10:00 AM	Wednesday
4	7	14	20	4000	Apr	11:00 AM	Thursday
5	9	18	25	5000	May	12:00 PM	Friday
6	11	22	30	6000	Jun	1:00 PM	Saturday
7	13	26	35	7000	Jul	2:00 PM	Sunday

Fig 4.16 Columns of different kinds of data users can autofilled

The columns are explained below.
- Odd numbers: When the first two numbers are entered, Excel knows that the step value is 2 and autofills the column with odd numbers.
- Every 4th number: In this example the step value is 4, so beginning with 2 and 6, Excel continues to increment each subsequent number by 4.
- Multiples of 5: The Fill Handle is handy for creating multiples of a number. Multiples are when each number can be divided evenly by the first number... which is also the step value in our example.
- Thousands: Excel autofills by thousands as step value is 1000.
- Months, time, and days: The last three columns illustrate Excel's capability of auto filling a variety of data series.

Sometimes users need autofill Linear Series and Skipping Rows. However, it is not uncommon to skip rows in a busy worksheet to enhance readability. To have Excel autofill every "nth" row with a linear series with a step value of 1, users have found that only one number usually needs to be entered, though Microsoft recommends always entering a minimum of two numbers. Be sure to always check the auto fill results for accuracy, see Fig 4.17.

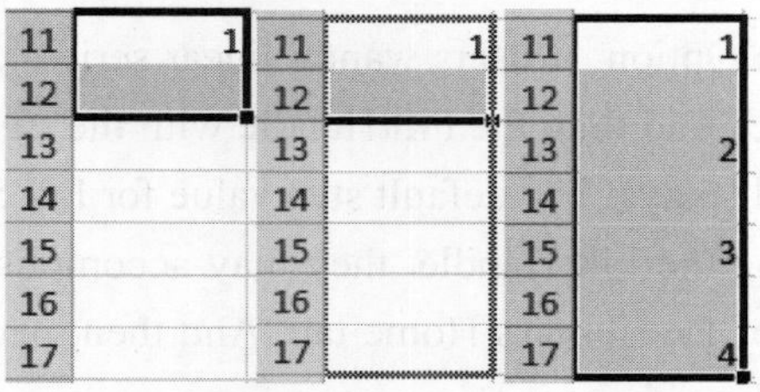

Fig 4.17 Using the fill handle to fill every other row in an Excel worksheet

The Fig 4.18 shows an example of autofilling numbers every other row. Enter the number 1, highlight that cell and the one below, and then click the Fill Handle with the left mouse button, drag, and release. Excel autofills every other row.

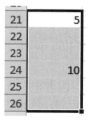

Fig 4.18   Autofill every third row

The fig also shows how people highlight for the Fill Handle to have Excel autofill every third row. If users were autofilling 1,2,3,4, etc., only the first number would need to be entered. But in this example users are autofilling with multiples of 5, so they must enter the first two numbers in their series so Excel can determine the step value. Then users must remember to highlight the appropriate number of blank cells, five in this example, after the second number.

In the former section the book introduced that each number in a linear series is calculated by adding the step value to the previous number. In a growth series, the next number is obtained by multiplying the previous number by the step value (or constant).

There are two methods for instructing Excel to autofill with a growth series or geometric pattern.

First, enter First Two Numbers in the Series.

Specifying growth series in newer versions of Excel is easy. The fastest way is to enter the first two numbers and then right-click on the Fill Handle.

The example is shown in Fig 4.19 has Excel autofill a worksheet column with a step value of 2.

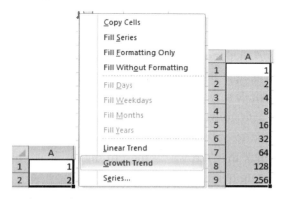

Fig 4.19   Using the fill handle1.

- In the Excel worksheet, enter 1 in the first cell and 2 in the cell immediately below.
- Highlight the two cells, and hover the cursor over the Fill Handle (+).
- Press the "right" mouse button, drag down the column, and release. The menu shown in fig

appears. Click "Growth Trend".
- Because the example entered two numbers, Excel knows that the step value is 2. The autofill results are shown.

If users want a step value of 3, enter 1 and 4 into the column. What if they want a step value of 2, but they want the series to begin with the number 3? Users would enter 3 and 6 because 3*2=6.

People must enter two numbers to use this quick method. If they only enter one number, the "Growth Trend" option is grayed out.

The second method entails an extra step, but it is nifty. The Fig 4.20 shows the process of creating a growth series beginning with the number 1 and having a step value of 3.
- Enter number 1 in the cell. Highlight the cell, click and hold on the "Fill Handle" with the right mouse button, and drag.
- When the mouse button is released a menu appears. Click "Series" at the bottom of the menu.
- The Series window opens. Type 3 in the Step value box and click "Growth" in the "Type" section. Click "OK".
- Excel fills the column as shown in the last image below.

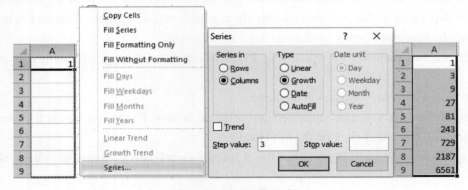

Fig 4.20 Step value of 3

Image (Fig 4.21) of worksheet column with 1 in the first cell, the column highlighted and the cursor hovering over the fill handle image of the menu that appears when right-clicking on the fill handle image of the Series window where we specify growth series and a step value of 3 image of the worksheet column filled with growth series.

There is always a method to open the Series Window by using the Ribbon. The image shown below appears when users click the "arrow" by the Fill button image of the Editing section of the Home tab with an arrow pointing to the Fill button.
- After the number 1 has been entered, click back in the cell and highlight it and the cells below it that people want autofilled.
- With the column highlighted, click the Fill button located on the Editing section of the Home tab as shown in the first image.

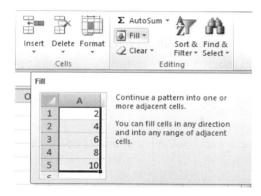

Fig 4.21   Autofill

- A drop-down menu appears as shown in the second image. Click Series and the Series window appears.

Sometimes, people need to autofilling Dates (Days, Weekdays, Months, and Years). They can instruct Excel to autofill cells with dates using the Fill Handle "right-click" function or via the Fill button on the ribbon. Below (Fig 4.22) is an example of using the Fill Handle to autofill dates.

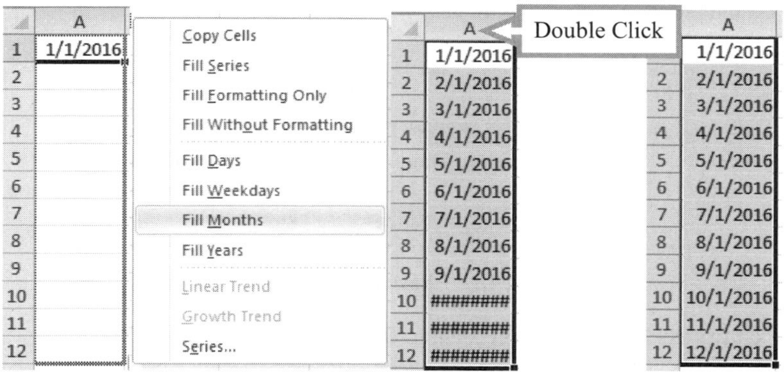

Fig 4.22   Autofill Dates

- Type the beginning date in the cell. For our example, people entered the first day of the year. Right-click on the Fill Handle and drag down the column.
- When the mouse button is released, a menu displays as shown in the first image. Users can select Months in the example.
- Excel autofills the column with the first day of each consecutive month as shown.
- When cells shows########, it means the width of the cell is not enough and users can adjust the cells to show entire date by double click the right vertical line.

The Fill button can also be used to autofill dates with more functionality. It not only creates a series based on the unit users select, but it increments that unit (days, months, etc.) by the number entered in the "Step value" box.

For example, after entering the first date and highlighting the column cells, click the "Fill" button and click "Series" from the drop-down menu that appears. When Dates is selected on the Series window (shown in the last section), the date unit is selected on the right side of the window. The step value default is 1, but users may select any value. For example, if people want every other month autofilled, enter a step value of 2.

4.2.3　Closing a Worksheet

Here are the steps to close a workbook. Firstly, users need to click the "Close" button as shown in Fig 4.23.

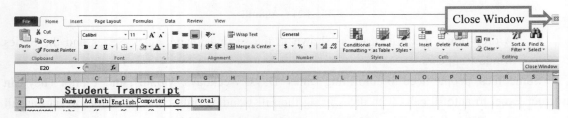

Fig 4.23　Closing a Worksheet

If the user has not save the table, he will see a confirmation message to save the workbook like Fig 4.24.

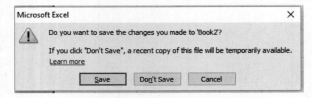

Fig 4.24　Save the workbook

And then he can press "Save" button to save the workbook.

4.3　Formulas and Functions

Before starting formulas and functions, the cell references need to be fully understood. In Excel formulas, people can refer to other cells either relatively or absolutely. When they copy and paste a formula in Excel, how to create the references within the formula tells Excel what to change in the formula it pastes. The formula can either change the references relative to the cell where people are pasting it (relative reference), or it can always refer to a specific cell. They can also mix relative and absolute references so that, when they move or copy a formula, the row changes but the column does not, or vice versa.

Preceding the row and/or column designators with a dollar sign ($) specifies an absolute reference in Excel shown as table 4.2.

TABLE 4.2 Absolute reference

Example	Comment
=A1	Complete relative reference
=$A1	The column is absolute; the row is relative
=A$1	The column is relative; the row is absolute
=$A$1	Complete absolute reference

The effect on cell references when copying formulas, Excel interprets cell references relationally. Because of this, when a formula is copied from one cell to another, Excel will change the cell names in order to keep the same relationships!

For example, if people copy the formula in cell A3 (=A1+A2) and paste it into cell B3, Excel will change the formula to read =B1+B2. And if the formula is pasted into cell F3, Excel will change the formula to =F1+F2 (see Fig 4.25 and Fig 4.26).

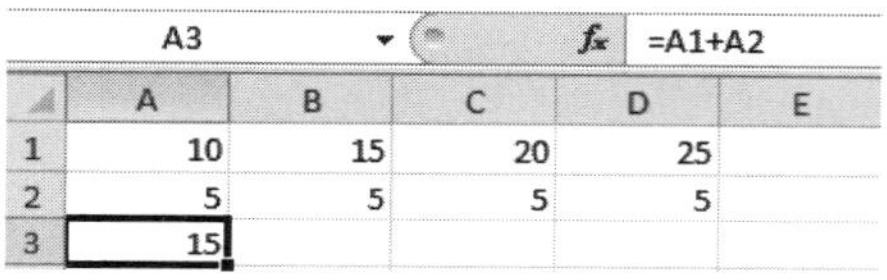

Fig 4.25 Formulas A1+A2

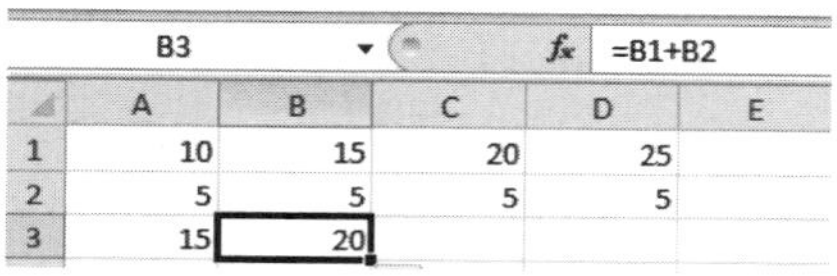

Fig 4.26 Copying Formulas to B1+B2

when people so often want to total or sub-total the data values in a series of rows or columns, they will find how useful it is!

When pasting the contents of a cell into multiple cells, just click on the cell containing the formula, copy it, and paste into the remaining cells—using the arrow keys on the keyboard to move from cell to cell.

What if people want to copy a formula but DON'T want Excel to automatically change one or more of the cell references? In these instances they will want to use an absolute cell reference in their formula.

In the sample spreadsheet below, people want to divide each number in Row 2 by the number in A1, and place the answers in Row 3. Normally, the formula in cell A3 would be =A2/A1. However, if people copy and paste this formula into B3, Excel will change the formula to =B2/B1 and this is not the result people need! They need the formula in cell B2 to read =B2/A1.

Excel worksheet example showing absolute cell reference. They tell Excel not to change A1 in the formula by using an absolute cell reference for A1. To specify an absolute cell reference, people can place a $ before the column letter and row number of the cell. The formula in cell A3 should now read =A2/$A$1. When copying and pasting the formula into other cells, Excel will keep A1 constant. Notice in the sample worksheet above that the formula in cell B3 shown in the formula bar is =B2/$A$1 (Fig 4.27 and Fig 4.28).

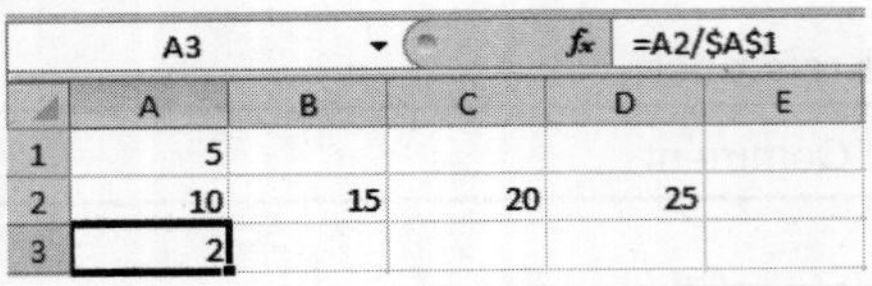

Fig 4.27 Absolute cell reference

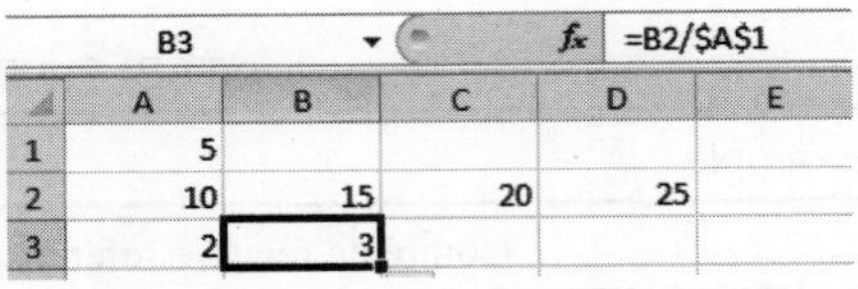

Fig 4.28 Absolute cell reference

4.3.1 Formulas

A formula is an expression which calculates the value of a cell. Functions are predefined formulas and are already available in Excel.

For example, cell A3 below contains a formula (Fig 4.29) which adds the value of cell A2 to the value of cell A1.

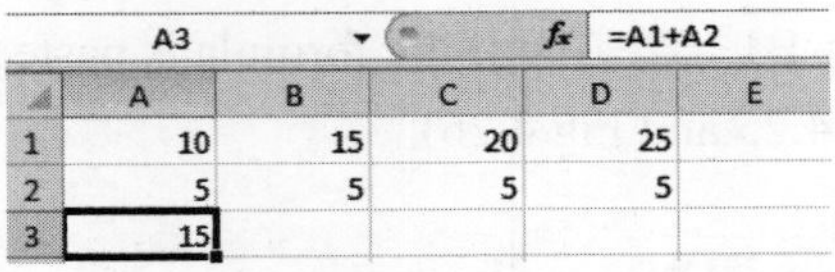

Fig 4.29 Formulas

To enter a formula, execute the following steps.
- Select a cell.
- To let Excel know that people want to enter a formula, type an equal sign (=).
- For example, type the formula A1+A2. (Fig 4.30)

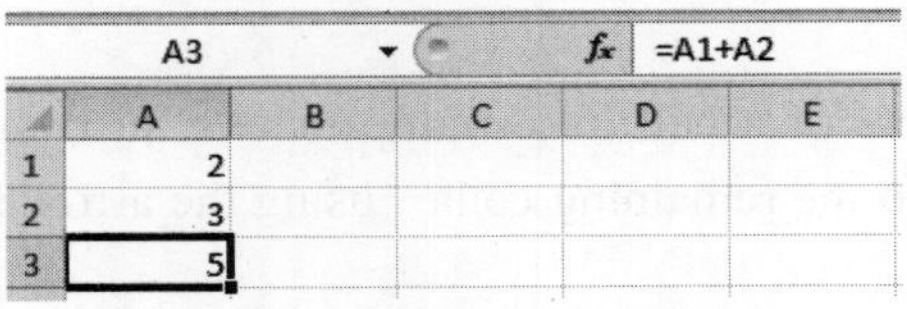

Fig 4.30 enter a formula

- Change the value of cell A1 to 3. (Fig 4.31)

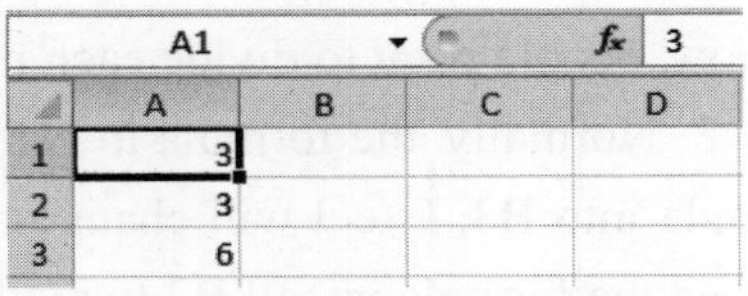

Fig 4.31 Change the value

Excel automatically recalculates the value of cell A3. This is one of Excel's most powerful features!

When users select a cell, Excel shows the value or formula of the cell in the formula bar.
- To edit a formula, click in the formula bar and change the formula.
- Press Enter.

Excel uses a default order in which calculations occur. If a part of the formula is in parentheses, that part will be calculated first. It then performs multiplication or division calculations. Once this is complete, Excel will add and subtract the remainder of formulas. See the example below.

First, Excel performs multiplication (A1 * A2). Next, Excel adds the value of cell A3 to this result. (Fig 4.32 & 4.33)

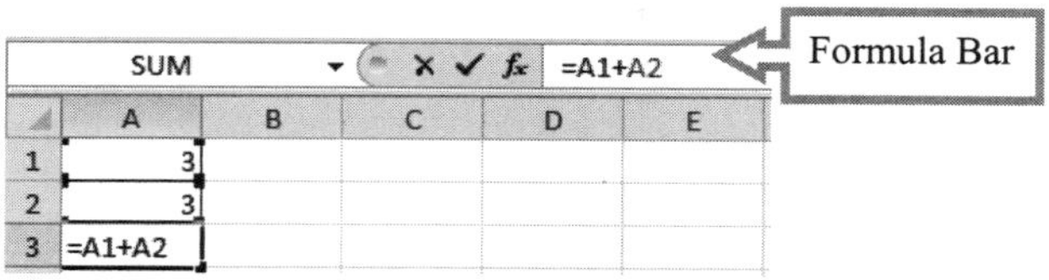

Fig 4.32   Edit a formula

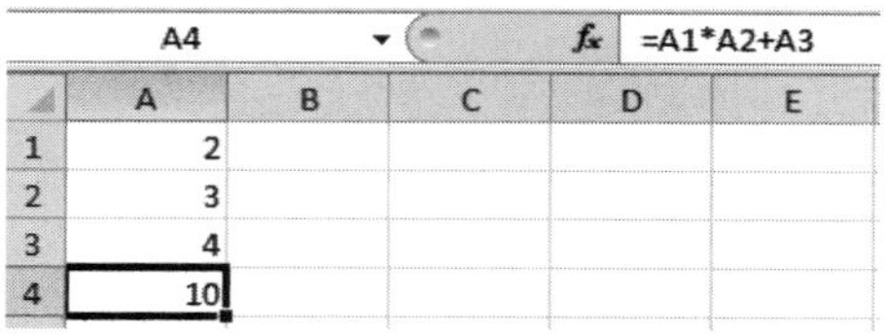

Fig 4.33   Formula =A1*A2+A3

In this case, Excel calculates the part in parentheses (A2+A3) first. And then, it multiplies this result by the value of cell A1. (Fig 4.34)

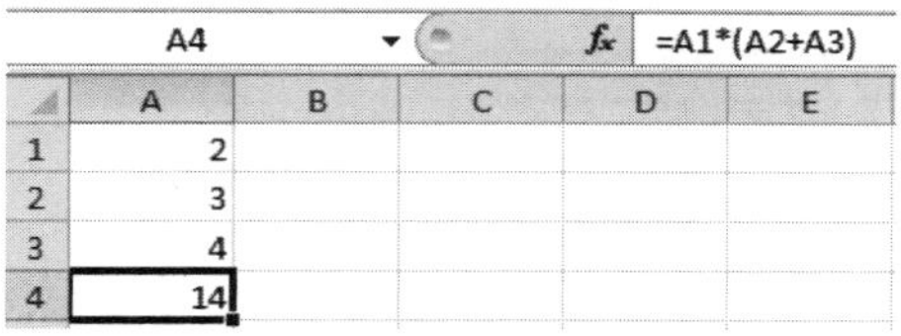

Fig 4.34   Change in parentheses

When people copy a formula, Excel automatically adjusts the cell references for each new cell the formula is copied to. To understand this, execute the following steps.

- Enter the formula shown in Fig 4.35 into cell A4.

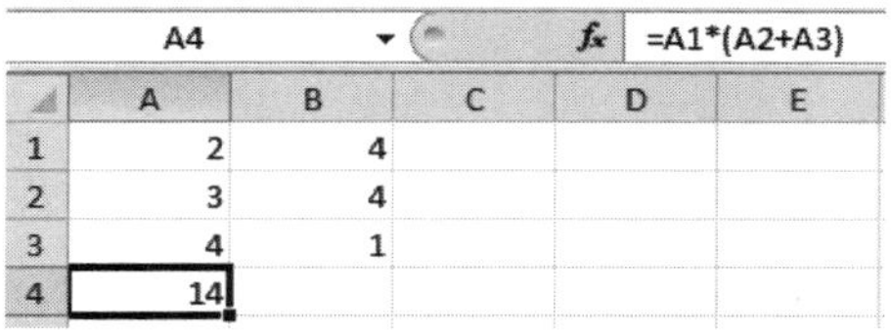

Fig 4.35   Result

- Select cell A4, right click, and then click "Copy" (or press CTRL + C). Next, select cell B4, right click, and then click Paste under "Paste Options" (or press CTRL + V).
- people can also drag the formula to cell B4. Select cell A4, click on the lower right corner of cell A4 and drag it across to cell B4. This is much easier and gives the exact same result! (Fig 4.37)

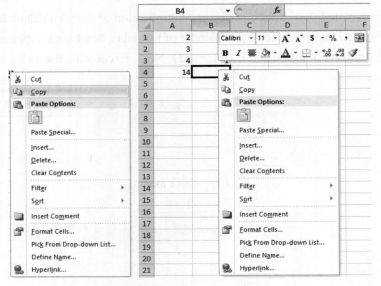

Fig 4.36   Copy the formula

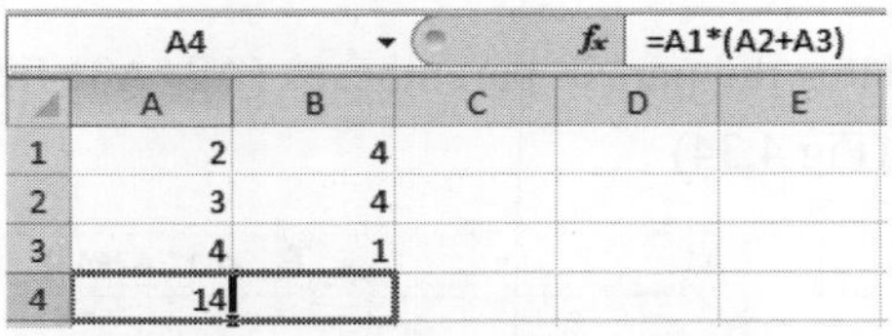

Fig 4.37   Drag the formula

Result. The formula in cell B4 references the values in column B. (Fig 4.38)

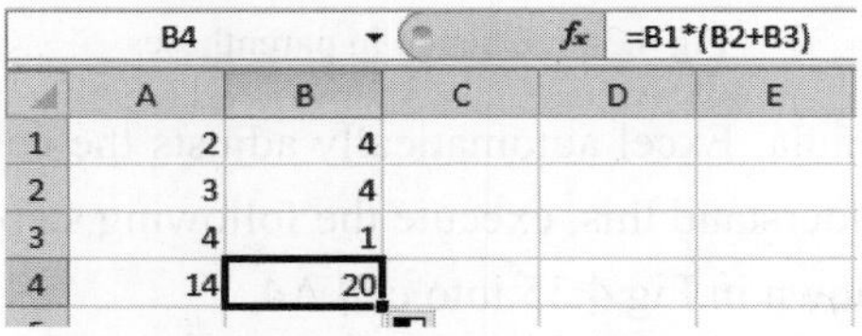

Fig 4.38   Result

4.3.2 Functions

A function in Excel calculates a result based on one or more input values. For example, the SUM function returns the sum of all the cells people specify as arguments. If a user was to type into a cell the following:

=SUM(B2:B6)

Then SUM is the function, B2:B6 are the arguments and the whole thing is a formula.

SUM is a straightforward function, but what if people want to use a function whose name don't know. Or perhaps users know the function's name, but don't know what arguments it needs. This is where the Insert Function command can help. The "Insert Function" button is located in the formula bar. The Insert Function dialog box in Excel 2010 simplifies the task of using functions in users' Excel worksheets. The Insert Function dialog box helps people locate the proper function for the task at hand and also provides information about the arguments that the function takes. If people use the Insert Function dialog box, they don't have to type functions directly into worksheet cells. Instead, the dialog box guides through a (mostly) point-and-click procedure.

Arguments are pieces of information that functions use to calculate and return a value.

- Display the Insert Function dialog box.

The most common ways to do this are by clicking the "Insert Function" button on the Formulas tab or by clicking the "Insert Function" button (which looks like "fx") on the Formula bar. Users can also access the Insert Function dialog box by clicking the small arrow at the bottom of the AutoSum button on the Formulas tab of the Ribbon and selecting More Functions (shown in Fig 4.39).

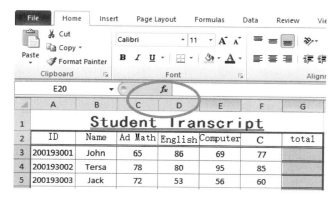

Fig 4.39   Function Dialog

- Select a function category in the Select a Category list.

If people don't know which category to choose, select "All" to display all functions in the Select a Function list. People also can type a brief description in the "Search for a Function" box and click "Go" like Fig 4.40.

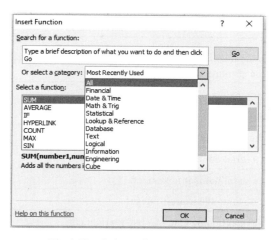

Fig 4.40   Select a function category

- Select the desired function in the "Select a Function" list box.

Description of the selected function, along with the function syntax, appears at the bottom of the dialog box.

- Click "OK".

The Function Arguments dialog box appears. This is where a user enters or selects the arguments for the function. Click the "Help" on this function link at the bottom of the dialog box for more details on the function.

- Enter the function arguments and click "OK".

Arguments may be references to cells, text or numbers that people type directly into the text boxes, or even other formulas. Click "Cancel" if they want to return to the worksheet without entering a function (like Fig 4.41).

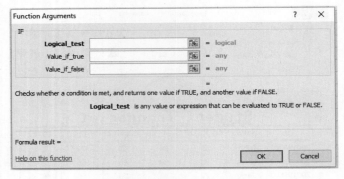

Fig 4.41　Function arguments

People can access and add functions directly from the Function Library on the Formulas tab of the Ribbon. Click the button representing the category of function people want (such as Financial) and select the desired function.

Excel 2010 comes with many built in functions that cover a wide range of topics. Some of the more commonly used ones are given below.

Table 4.2　Some functions

Name	Functions
ABS	Returns the absolute value of a number
AVERAGE	Adds its arguments
COUNT	Counts how many numbers are in the list of arguments
Date	Returns the serial number of a particular date
Day	Converts a serial number to a day of the month
Hour	Converts a serial number to an hour
Rounds	a number down to the nearest integer
MAX	Returns the maximum value in a list of arguments
MIN	Returns the minimum value in a list of arguments

续表

Name	Functions
Minute	Converts a serial number to a minute
Month	Converts a serial number to a month
Now	Returns the serial number of the current date and time
SUM	Adds its arguments
IF	Test a condition and have one value returned if the condition is TRUE, and another value returned if the condition is FALSE
COUNTIF	Calculate the number of cells in a single range whose values meet specific criteria
Today	Returns the serial number of today's date

There are some typical functions need to be introduced.

- Average():AVERAGE(number1, number2, number3...)

In this function number1, number2, etc. are from 1 to 255 arguments for which Excel should find the average. The arguments can be numbers; or cell references, cell ranges, formulas, or other functions that resolve to a number.

Cells that contain the AVERAGE function should be formatted to a desired number of decimal points as the decimal portion may be infinite when averaging numbers. For example:

=AVERAGE(4,5,2) ... This Excel function adds three numbers and divides by 3: (4 + 5 + 2 = 11; 11 ÷ 3 = 3.66666... If the cell is formatted to display 2 decimal points, Excel returns a value of 3.67.

In the worksheet examples, column A contains the actual function, so what it shown is the value Excel has returned. The text of the function is displayed in column B. The data resides in cells C1 through C4 (Shown in Fig 4.42).

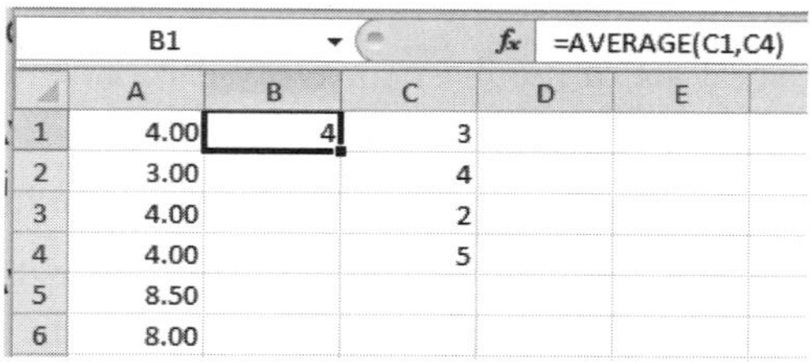

Fig 4.42 An Example of Function average()

Example B1: =AVERAGE(C1,C4). This example shows that individual cell references can be arguments. The average of C1 and C4 is 4: 3 + 5 = 8; 8 ÷ 2 = 4.

Example B2: =AVERAGE(C1:C3). This example shows the function containing a cell range. The average of the first 3 cells in Column C is 3: 3 + 4 + 2 = 9; 9 ÷ 3 = 3.

Example B3: =AVERAGE(C1:C2,5). This example shows that arguments can be numbers: 3 + 4 + 5 = 12; 12 ÷ 3 = 4.

Example B4: =AVERAGE(C1:C2,C4). This example mixes a cell range with a cell reference. Since C4 contains the number 5, the answer is the same as B3.

Example B5. =AVERAGE(PRODUCT(C1,C2),C4). This example shows an embedded function. This formula takes the average of 12 (product of 3 x 4) and 5 to arrive at a value of 8.5.

Example B6: =SUM(AVERAGE(C1:C3),C4). This example shows the AVERAGE function embedded in a SUM function. This formula adds the average of C1 through C3, or 3, to the contents of cell C4, which is 5, for a value of 8.

- If():IF(logical-test, value-if-true, value-if-false) logical-test is a condition which must evaluate to either TRUE or FALSE: value-if-true is the value Excel returns if the logical test evaluates to TRUE; value-if-false is the value that Excel returns if the logical test evaluates to FALSE.

There are some examples of the IF Function:

=IF(D1>26,33,44) - This Excel function checks to see if the value in D1 is greater than 26. If so, Excel returns a value of 33. If not, a value of 44 is returned.

=IF(A5<>, "Done" "Open"): This Excel function, a value of Done will be returned if cell A5 is not empty. Otherwise, a value of Open will be returned.

=IF(D1<100, "OK" "Over Budget"): This Excel formula checks to see if the value in D1 is less than 100. If so, Excel displays OK. If not, Excel displays Over Budget. Text must be wrapped in quotation marks.

=IF(SUM(A1:A3)=50, "Slow" "Fast"): This Excel formula checks to see if the sum of the contents of cells A1 through A3 equals 50. If so, Excel returns a value of Slow. If not, Fast is displayed. For more information about the SUM function, see our tutorial The SUM Function.

=IF(A2>1000,985,B2+245): This Excel formula checks to see if the value in A2 is greater than 1000. If so, Excel returns a value of 985. If not, Excel displays the sum of B2 and 245.

- SUM():The Excel SUM function adds together a supplied set of numbers and returns the sum of these values. The syntax of the function is: SUM(number1, number2...) where the number arguments are a set of numbers (or arrays of numbers) that users want to find the sum of.

Excel SUM Function Calculation are Included in Numbers and dates, which are always counted as numeric values by the Excel SUM function. However, text representations and logical values are handled differently, depending on whether they are values stored in the cells of spreadsheets, or they are supplied directly to the function.

The table below summarizes which values are included in the Excel SUM Function calculation, and which values are ignored or result in errors:

Table 4.4 Sun() Contents

	Value Within a Range of Cells	Value Supplied Directly to Function
Numbers	Included	Included
Dates	Included	Included
Logical Values	Ignored	Included (True=1; False=0)

	Value Within a Range of Cells	Value Supplied Directly to Function
Text Representations of Numbers & Dates	Ignored	Included
Other Text	Ignored	#VALUE! Error
Errors	Error	Error

There are Sum Function Examples in the following spreadsheet, the Excel SUM function is used to calculate the sum of the numbers 5, 6, 7, 8, 9. In each of the five examples, the numbers are supplied to the Sum function in a different way (see Fig 4.43).

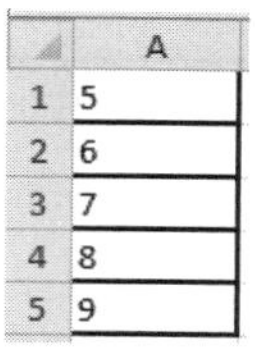

	A
1	5
2	6
3	7
4	8
5	9

Fig 4.43　An Example of Sum()

Table 4.5　Sum() Example

Formulas:	Results:
=SUM(5, 6, 7, 8, 9)	35
=SUM({5,6,7}, 8, 9)	35
=SUM(A1, A2, A3, A4, A5)	35
=SUM(A1, A2, A3, "8" "9")	35
=SUM(A1:A5)	35

Note that, in the above example spreadsheets:

Each argument to the Sum Function can be supplied as a single value or cell, or as an array of values or cells (note that in cell B2, the argument {5,6,7} is an array of numbers); and when supplied directly to the function, text representations of numbers are included in the Sum calculation (see the example in cell B4).

- ROUND():Round Function rounds a supplied number up or down, to a specified number of decimal places. The syntax of the function is ROUND(number, num_digits).

Where the arguments are number (The initial number) and num_digits(The number of decimal places to round the supplied number to).

In this function a positive num_digits value specifies the number of digits to the right of the decimal point; num_digits value of 0 specifies rounding to the nearest integer; A negative num_digits value specifies the number of digits to the left of the decimal point.

There are some examples. (Fig 4.44)

Column B of the following spreadsheet shows several examples of the Excel Round function:

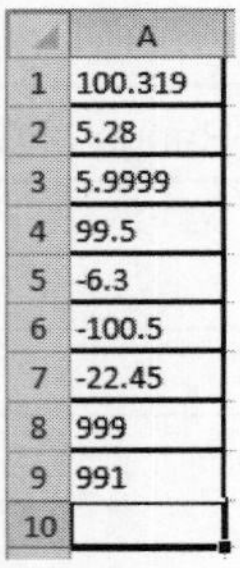

	A
1	100.319
2	5.28
3	5.9999
4	99.5
5	-6.3
6	-100.5
7	-22.45
8	999
9	991
10	

Fig 4.44   An Example of Round()

Table 4.6   Round Examples

Formulas:	Results:
=ROUND(A1, 1)	100.3
=ROUND(A2, 1)	5.3
=ROUND(A3, 3)	6
=ROUND(A4, 0)	100
=ROUND(A5, 0)	-6
=ROUND(A6, 0)	-101
=ROUND(A7, 1)	-22.5
=ROUND(A8, -1)	1000
=ROUND(A9, -1)	990

The above examples show how the Round Function rounds up or down, to the specified number of decimal places.

It is often a good idea to use the Round Function when comparing two numbers in Excel, especially if the numbers are the result of multiple mathematical calculations. This is because multiple calculations may result in the introduction of rounding errors, which may cause small inaccuracies in the numbers stored in Excel.

For example, the decimal 5.1 may, due to rounding errors, be stored in cell A1 of spreadsheet, as 5.10000000000001. When compared to the exact value 5.1, the value in cell A1 will not be equal to the exact value 5.1.

However, applying the Round Function to the value in cell A1 removes the rounding error.

I.e. the expression is ROUND(A1, 1) = 5.1 or even the expression is ROUND(A1, 10) = 5.1 will evaluate to TRUE.

- COUNT():The COUNT Function calculates the number of cells that contain numeric values or the number of arguments in the function, if any, that are numeric. The syntax of the COUNT function is COUNT(value1, value2...). The value1 is required and is an item, cell reference, or range of cells, and value2... are optional and are 1 to 255 additional items, cell references, or ranges. For example, the function =COUNT(5,10,15) would ask Excel to count the number of arguments that are numeric (3). And the function COUNT(C1:C22) would ask Excel to count the number of cells in the range C1 through C22 that contained a numeric value. Excel counts as a numeric value any number, currency, date, time, and percentage. Excel also counts numbers that reside in a cell that has been formatted for "text", as long as the cell only contains the number. For example, Excel would not count "50 cents" when using the COUNT Function.

There are examples of the COUNT Function:

In the spreadsheet below, columns A and B contain data. Column C contains the actual COUNT Function, so what we see is the function's result. Column D shows the function. Follow along as we discuss each of the three examples below the worksheet image.

	A	B
1	Cost	Sales
2	6.89	58
3	2/3/2016	Good
4		9/18/2004
5	#N/A	10

Fig 4.45   An Example of Count()

Table 4.7   Count() Examples

Formulas	Results
=COUNT(A1:A5)	2
=COUNT(A1:A5,10)	3
=COUNT(A1:A5,B1:B5)	5

=COUNT(A1:A5): This Excel Function counts the number of cells in the range A1 through A5 that contain a numeric value. Excel returns a value of 2(cells A2, A3). Cells A1,A4 and A5 are not counted because they do not contain numbers.

=COUNT(A1:A5,10): The number of cells with numeric values in cells in A1 to A5 is 2, plus 1 for the number 10; so Excel displays 4 (cells A2, A3, and the number 10).

=COUNT(A1:A5,B1:B5): This Excel function counts all cells containing numeric values in cells A1 to A5, and cells B1 to B5. Excel returns a value of 5 (cells A2, A3, A5, B2, and B4).

4.4   Formatting a Worksheet

Each cell in a worksheet can be formatted with many properties. However, people must point out here that the format of a cell does not affect the actual value in the cell. If a cell contains a math formula, especially division, multiplying of decimals, or one containing complex calculations, the actual value that Excel computes may have a large number of decimal places. But if the cell has been formatted for 2 decimal points, that are entire Excel will display. But depending on the situation, the value people see on the worksheet may not be the actual value for the cell.

There are several tabs in the "Format Cells" window, and all formatting options may be found on one of these tabs. Multiple cells can be formatted in one step by first selecting the cells.

The "Format Cells" window can be accessed in all versions of Excel from the right-click menu. Since Microsoft seems to change the user interface so often, users feel it is faster and easier to just right-click to reach the Format Cells window.

Some formatting options are available on the Font, Alignment, and Number groups of the Home tab. To see formatting options not displayed, click the little arrow in the lower right corner of the group and the "Format Cells" window displays.

4.4.1 Editing the cells

MS Excel Cell can hold different types of data like Numbers, Currency, Dates, etc.. People can set the cell type in various ways. Therefore, cell formatting is one important rule for users.

- Cell formatting: people can right click on the cell→Format cells→Number or they can click on the Ribbon from the ribbon see Fig 4.46.

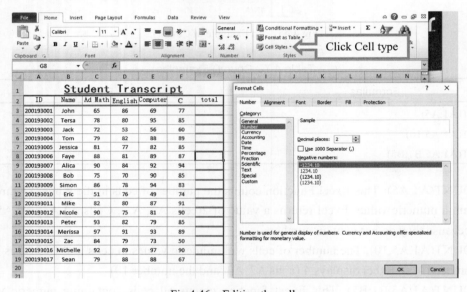

Fig 4.46　Editing the cells

There are various Cell formats, and shown below:

General: This is default cell format of Cell.

Number: This displays cell as number with separator.

Currency: This displays cell as currency i.e with currency sign.

Accounting: Similar to Currency used for accounting purpose.

Date: Various date formats are available under this like 17-09-2013, 17th-Sep-2013, etc.

Time: Various Time formats are available under this like 1.30PM, 13.30, etc.

Percentage: This displays cell as percentage with decimal places like 50.00%.

Fraction: This displays cell as fraction like 1/4 ,1/2 etc.

Scientific: This displays cell as exponential like 5.6E+01.

Text: This displays cell as normal text.

Special: This is special formats of cell like Zip code, Phone Number.

Custom: People can use custom format by using this.

- Font format: Mainly, people can make additional font-related changes by using the Font tab in the Format Cells dialog box. To access these settings, click the Font dialog box launcher (that little icon in the lower-right corner of the Font group on the Home tab).

People can be changing the font or font attributes by select the cells they want to format. On the Home tab, open the Font drop-down menu and select a font.

As users hover their mouse over a font face, Excel's Live Preview feature displays the selected cells in the different fonts, see Fig 4.47.

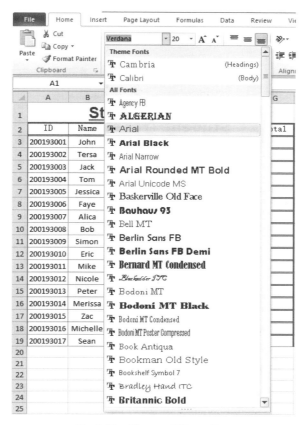

Fig 4.47   Change different font

In the Font group of the Home tab, click the "Font Size" arrow and select a font size. People can choose another size besides those listed by simply typing the desired font size in the Font Size box. If people want to quickly go up or down a font size, click the "Increase Font Size" or "Decrease Font Size" button in the Font group.

People also can click a button to apply a font attribute in the Font group of the Home tab.

The Bold, Italic and Underline buttons are toggle switches — click the button once to turn the attribute on, and click it again to turn it off: Bold (or "Ctrl+B"), Italic (or "Ctrl+I"), and Underline (or "Ctrl+U").

The default underline style is a single underline. Click the "Underline" button down arrow to choose Double Underline. Or, choose additional underline options through the Font dialog box launcher, which displays the Format Cells dialog box. Underline options appear on the Font tab.

- People can change the color of a font, follow these steps: Select the cells they want to format. Then, in the Font group of the Home tab, click the "Font Color" drop-down arrow and select a color. If users hover the mouse over a color, Excel's Live Preview feature shows the selected cells in that font color. Click "More Colors" to display the Colors dialog box from which users can select additional colors as well as create own custom color, like Fig 4.48.

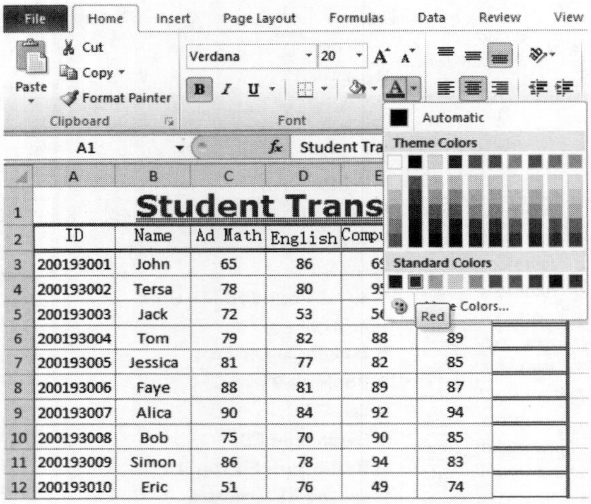

Fig 4.48 Change different color

- Merging and centering cells

Users can merge and center data horizontally or vertically across multiple cells in Excel 2010. People also can unmerge or split a merged cell into its original, individual cells. A common use of merge and center in Excel 2010 is to horizontally center a worksheet title over a table.

People could select the range of cells they want to merge and center. They can use Merge & Center only on a contiguous, rectangle-shaped range of cells (Fig 4.49).

On the Home tab, in the Alignment group, click the Merge & Center button. The cells are merged into a single cell, and the text (if any) is centered within the merged cell.

However, sometimes people need to split a merged cell. If people need to split a cell that they've merged with the "Merge & Center" button, follow these steps: Select the merged cell by the "Merge & Center" button appears selected in the Alignment group. And then, people can click the "Merge & Center" button in the Alignment group. The merged cell reverts to a cell range again, and any text contained in the merged cell appears in the upper-left cell of the range.

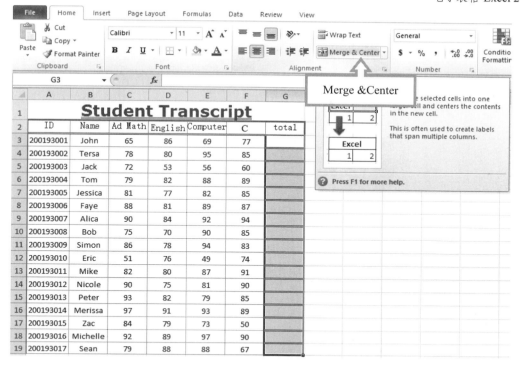

Fig 4.49   Merge the cells

4.4.2 Borders and Shadings

In Excel 2010, users can add borders to individual cells to emphasize or define sections of a worksheet or table. Use the Borders button in the Font group on the Home tab to add borders of varying styles and colors to any or all sides of the cell selection.

The following steps can help people add cell borders: Select the cells people want to format. And then click the "down arrow" beside the Borders button in the Font group on the Home tab. A drop-down menu appears, with all the border options users can apply to the cell selection (Fig 4.50).

Last, click the type of line users want to apply to the selected cells.

When selecting an option on the Borders drop-down menu, people need to be aware of that Excel draw borders only around the outside edges of the entire cell selection (in other words, following the path of the expanded cell cursor), click the "Outside Borders" or the Thick Box Border option. Or if users want borderlines to appear around all four edges of each cell they've selected, select the "All Borders" option. To change the type of line or line thickness or color of the borders people apply to a cell selection, open the Format Cells dialog box and use the options on its Border tab (click "More Borders" at the bottom of the Borders button's drop-down menu or press Ctrl+1 and then click the "Border tab") see Fig 4.51.

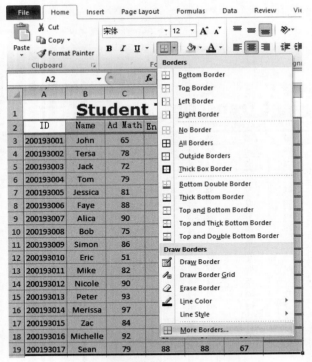

Fig 4.50   Borders and Shadings

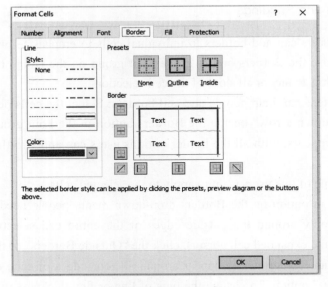

Fig 4.51   Change borders

Last but not least, to remove borders in a worksheet, select the cell or cells that presently contain them and then click the "No Border" option in the Borders button's drop-down menu.

Like Borders, to apply Shading, users can add shading to the cell from the Home tab→Font Group→Select the Color see Fig 4.52.

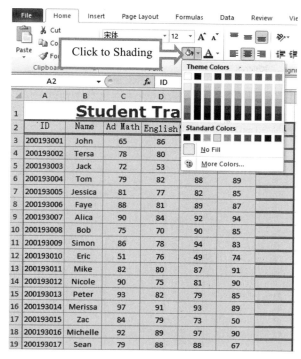

Fig 4.52　Change shading

4.5　Finalizing a Worksheet

In Excel 2010, it provides many ways to finalize data. For example, data sorting, filtering and worksheet print out. Like Word, Excel also can contain headers and footers. All the information will be introduce in this part.

4.5.1　Sorting data

Sorting data in MS Excel rearranges the rows based on the contents of a particular column. Users may want to sort a table to put names in alphabetical order. Or, maybe they want to sort data by Amount from smallest to largest or largest to smallest.

To sort the data is following steps like select the Column by which users want to sort data and then choose Data Tab→Sort. The Sorting dialog will appear (like Fig 4.53).

If users want sort data based on selected column Choose Continue with the selection or If they want sorting based on other columns choose Expand Selection, see Fig 4.54.

Sorting data can base on some conditions, which show below:

- Values: Alphabetically or numerically.
- Cell Color: Based on Color of Cell.
- Font Color: Based on Font color.
- Cell Icon: Based on Cell Icon.

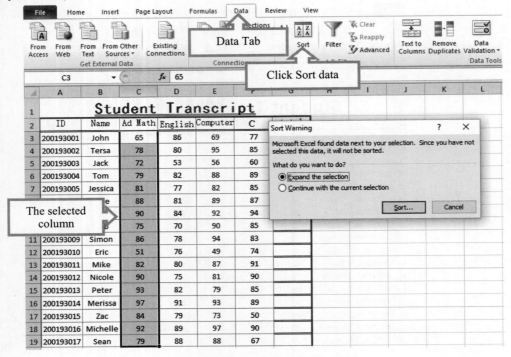

Fig 4.53 Sorting data

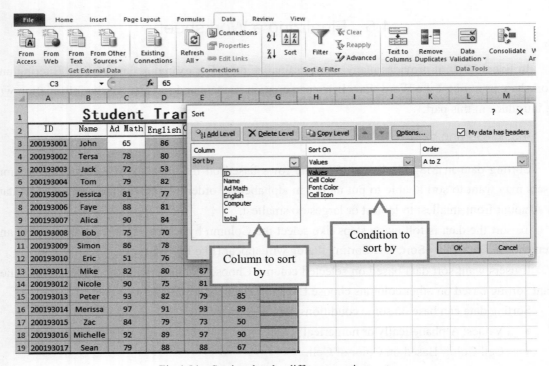

Fig 4.54 Sorting data by different requirements

Sorting option (Fig 4.55) is also available from the Home Tab. People can choose Home Tab→

Sort & Filter. Users can see same dialog to sort records.

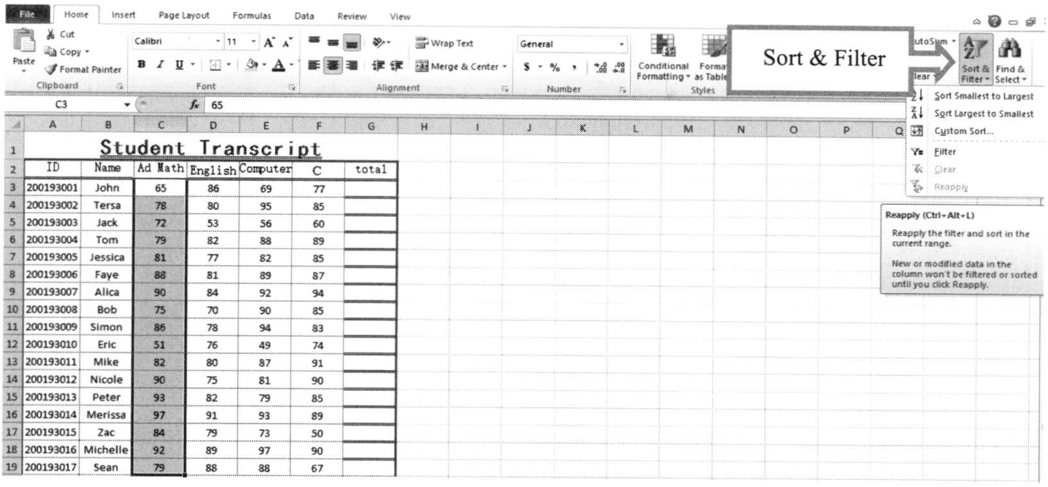

![Sorting option screenshot]

Fig 4.55　Sorting option

4.5.2　Filtering data

Filtering data in Excel refers to displaying only the rows that meet certain conditions. (The other rows get hidden.)

Using the store data, if an administrator is interested in seeing data where C course is under 60, then he can set filter to do this. Follow below steps to do this: place a cursor on the Header Row, and then choose Data Tab→Filter to set filter see Fig 4.56.

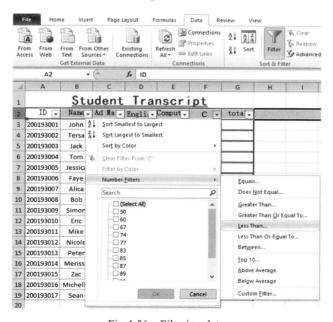

Fig 4.56　Filtering data

Click the "drop-down" arrow in the Area Row Header and remove the check mark from "Select All" which unselects everything. Then select the Number Filter "Less than 60" which will filter the data and displays data of under 60, like Fig 4.57.

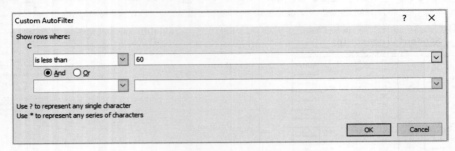

Fig 4.57   Custom Autofilter dialog

Users also can use Multiple Filters. They can filter the records by multiple conditions i.e. by multiple column values. Suppose after C course under 60 is filtered an administrator need to have filter where Ad Math is equal to 50. After setting filter for C course, choose Ad Math column and then set filter for 50 see Fig 4.58.

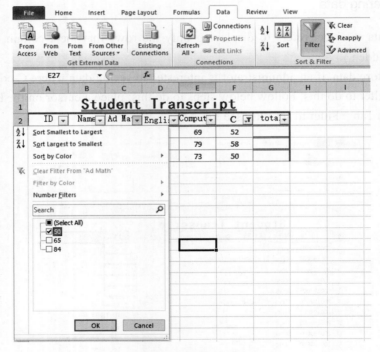

Fig 4.58   Multiple Filters

4.5.3   Headers and Footers

A header is information that appears at the top of each printed page and footer is information that appears at the bottom of each printed page. By default, new workbooks do not have headers or footers.

If people need to add Header and Footer, they can choose Page Setup dialog box→Header or Footer tab see Fig 4.59.

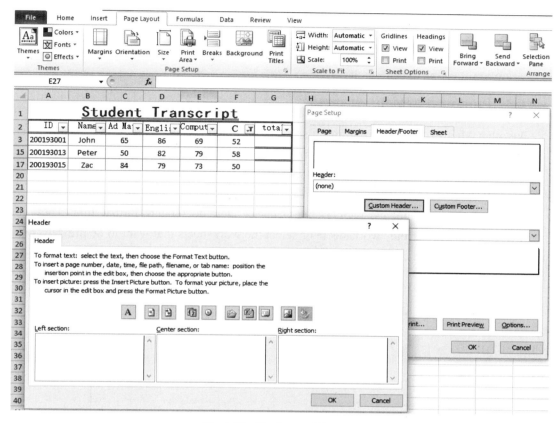

Fig 4.59 Headers and Footers

They can choose predefined header and footer or create own custom.
- &[Page]: Displays the page number.
- &[Pages]: Displays the total number of pages to be printed.
- &[Date]: Displays the current date.
- &[Time]: Displays the current time.
- &[Path]&[File]: Displays the workbook's complete path and filename.
- &[File]: Displays the workbook name.
- &[Tab]: Displays the sheet's name.

There are some other header and footer options. When a header or footer is selected in Page Layout view, the Header & Footer→Design→Options group contains controls that let users specify other options:
- Different First Page: Check this to specify a different header or footer for the first printed page.

- Different Odd & Even Pages: Check this to specify a different header or footer for odd and even pages.
- Scale with Document: If checked, the font size in the header and footer will be sized. Accordingly if the document is scaled when printed. This option is enabled, by default.
- Align with Page Margins: If checked, the left header and footer will be aligned with the left margin, and the right header and footer will be aligned with the right margin. This option is enabled, by default.

4.5.4 Preview and Print

Like other Microsoft Office programs, users can print and preview files from the same location by clicking "File" and then "Print". In some programs, like Excel and Word, Print Preview appears on the main Print screen. In this part, it will illustrate more information about printing and previewing from specific programs, see Fig 4.60.

Before print the documents, people should check their documents format first. They could preview their files by click the "worksheet" or select the worksheets that want to be previewed. And then Click "File", and then click "Print". From Keyboard shortcut users can also press "CTRL+F2".

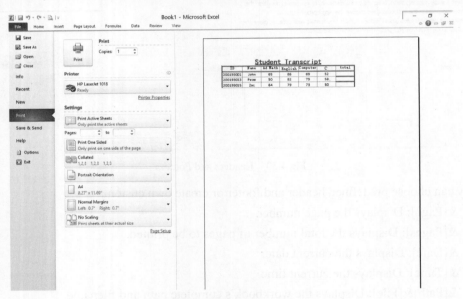

Fig 4.60 Preview and Print

To preview the next and previous pages, at the bottom of the Print Preview window, click "Next Page" and "Previous Page". If people need exit print preview and return to the workbook, click any other tab above the preview window.

When users print a partial or entire worksheet or workbook, they can do one of the following:
- To print a partial worksheet, click the "worksheet", and then select the range of data that want to be print.
- To print the entire worksheet, click the "worksheet" to activate it.

- To print a workbook, click any of its worksheets.

At last, users click "File", and then click "Print". From Keyboard shortcut users can also press "CTRL+P".

4.6 Creating a chart

A chart is a visual representation of numeric values. Charts (also known as graphs) have been an integral part of spreadsheets. Charts generated by early spreadsheet products were quite crude, but they have improved significantly over the years. Excel provides users with the tools to create a wide variety of highly customizable charts. Displaying data in a well-conceived chart can make numbers more understandable. Because a chart presents a picture, charts are particularly useful for summarizing a series of numbers and their interrelationships.

There are various chart types available in Excel as shown in below, see Fig 4.61.

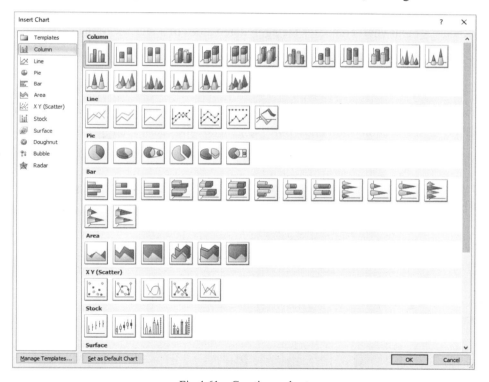

Fig 4.61   Creating a chart

- Column: Column chart shows data changes over a period of time or illustrates comparisons among items.
- Bar: A bar chart illustrates comparisons among individual items.
- Pie: A pie chart shows the size of items that make up a data series, proportional to the sum of the items. It always shows only one data series and is useful when users want to emphasize a significant element in the data.

- Line: A line chart shows trends in data at equal intervals.
- Area: An area chart emphasizes the magnitude of change over time.
- X Y Scatter: An xy (scatter) chart shows the relationships among the numeric values in several data series, or plots two groups of numbers as one series of xy coordinates.
- Stock: This chart type is most often used for stock price data, but can also be used for scientific data (for example, to indicate temperature changes).
- Surface: A surface chart is useful when people want to find optimum combinations between two sets of data. As in a topographic map, colors and patterns indicate areas that are in the same range of values.
- Doughnut: Like a pie chart, a doughnut chart shows the relationship of parts to a whole; however, it can contain more than one data series.
- Bubble: Data that is arranged in columns on a worksheet so that x values are listed in the first column and corresponding y values and bubble size values are listed in adjacent columns, can be plotted in a bubble chart.
- Radar: A radar chart compares the aggregate values of a number of data series.

4.6.1 Creating a new chart

To create charts for the data people select the data for which users want to create chart. They choose Insert Tab→Select the chart or click on the Chart group to see various chart types. Then, select the chart of users' choice and click "OK" to generate the chart, see Fig 4.62.

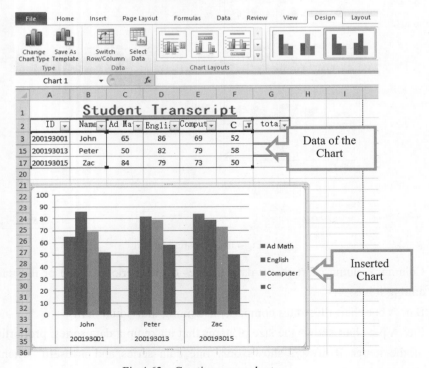

Fig 4.62　Creating a new chart

4.6.2 Modifying a chart

Users can modify the chart at any time after they have created it. Firstly, users can select the different data for chart input with right click on chart→Select data. Selecting new data will generate the chart as per new data as shown in the below screen-shot see Fig 4.63.

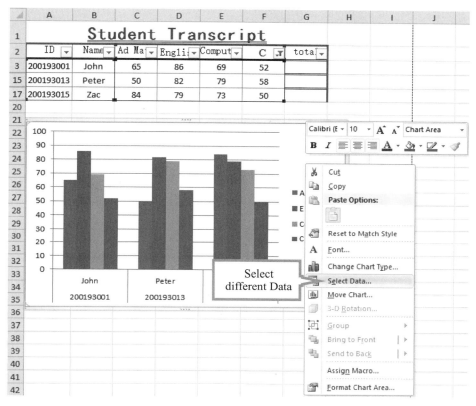

Fig 4.63　Modifying a chart

Users can change the X axis of the chart by giving different input to X-axis of chart. And users can change the Y axis of chart by giving different input to Y-axis of chart.

4.6.3　Formatting and editing a chart

Users may format or edit a chart after they finish insert chart. They can Delete, Move, and Resize Charts for their new requirements.

To select an existing chart, click on its border, or click in an empty space inside the chart. When selecting a chart, be careful not to click on an element inside the chart or that element will be selected instead.

- To delete a chart that has just been created, click the "Excel Undo" button. To delete an existing chart, select the chart and press the "Delete" key, or right-click and select "Cut".
- To move a chart to a different place on the worksheet, select the chart and drag it to the

desired location.
- To move a chart to a new or different spreadsheet in the same workbook, select the chart, right-click, and select Move Chart. Then choose the sheet or type in a new sheet name, and click "OK".
- To resize a chart, select the chart and drag any of the chart's corners.

Chapter 5   Microsoft PowerPoint 2010
电子幻灯片的制作——PowerPoint 2010

PowerPoint is a system in the Microsoft Office Suite that enables users to present information in office meetings, lectures and seminars to create maximum impact in a minimal amount of time. PowerPoint presentations can amplify people's message, accelerate the information being absorbed and assist with comprehension enabling faster decision making.

5.1 An overview of PowerPoint 2010

For anyone new to PowerPoint, it is always a good practice to get accustomed to the parts of the screen. For those of users who got on board with PowerPoint 2007, this screen will look very familiar. However, there are some new additions to PowerPoint 2010 in terms of features, and some subtle additions in terms of slight changes to existing features in PowerPoint 2007.

5.1.1 PowerPoint 2010 Components

Similar with Word, there are 2 ways to start PowerPoint: double click "PowerPoint documents" or navigate through the program by click "Start" button→"All apps"→"Microsoft Office"→"Microsoft PowerPoint 2010". Then, one PowerPoint window will show on the screen see Fig 5.1.

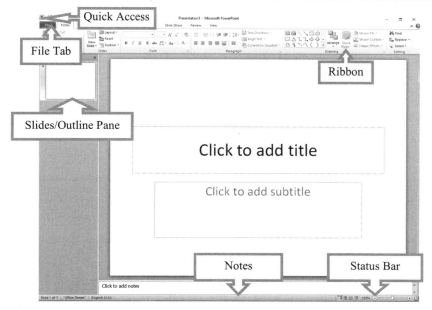

Fig 5.1   PowerPoint Windows

- File Tab: The new File tab in the left corner of the ribbon, replaces the Office button. Many of the same features are present and some new features have been added.
- Ribbon: The ribbon replaces the toolbar in older versions of PowerPoint, prior to PowerPoint 2007.
- Quick Access Toolbar: This toolbar appears in the top left corner of the PowerPoint 2010 screen. This is a customizable toolbar, so that users may add icons for features that they use frequently.
- Tabs on the Ribbon: These tabs on the ribbon are headings for groups of tasks. These tabs look similar to the headings on the menus in older versions of PowerPoint.
- Help Button: This tiny question mark icon is how to access help for PowerPoint 2010.
- Slides/Outline Pane: The Slides/Outline pane is located on the left side of the window. The Slides pane shows thumbnail versions of each of the slides in the presentation. The Outline pane shows a text outline of all the information on the slides.
- Notes: The Notes section is a place for the speaker to jot down any hints or references for his presentation. Only the presenter will see these notes.
- Status Bar: The Status bar shows current aspects of the presentation, such as the current slide number and what design theme was used. A tiny Common tools toolbar gives quick access to features that the presenter would use often.

5.1.2 Slides views

The different display modes of Powerpoint are normal mode, Slide Sorter mode, Reading view Mode, and Slide show mode. Each different mode can be accessed via the View Tab of the ribbon.
- The normal mode is the main editing mode, used to write and design people's presentations. Normal mode has four work areas:
  - The Outline tab is the location where users write the content (shows how texts used in their slides are to be displayed).
  - The Slides tab is the appropriate location to view the presentation slides as thumbnails while editing them. Thumbnails allow easy navigation for the presentation and visualization effects when changing the design. People can also rearrange, add or delete slides easily in this mode.
  - The main workspace, located in the upper right corner of the PowerPoint window, displays the user's active slide. People can add text and insert images, group/ungroup and manipulate as many as objects as they want: charts, SmartArt graphics, charts, drawing objects, the text boxes, movies, sounds, hyperlinks and animations.
  - The Comments Pane: below the Slide pane is used to enter comments about the slide. People can print the comments or refer to them during their presentation.

People can get into the Normal view in three methods: Normal View in menus if people have Classic Menu for Office, in Ribbon, or in Status bar. The fig 5.2 shows above are getting normal mode in Status bar.

Chapter 5　Microsoft PowerPoint 2010
电子幻灯片的制作——PowerPoint 2010

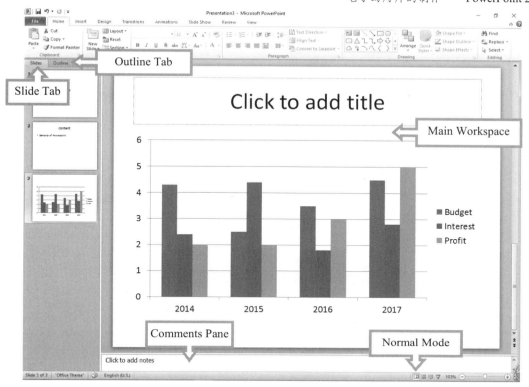

Fig 5.2　The normal mode

The Fig 5.3 below shows how to get into Normal View from Ribbon, if people do not have Classic Menu for Office. They need click the "View" tab, go to the Presentation Views group, and click the "Normal" button.

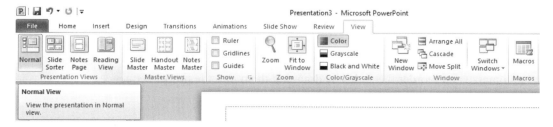

Fig 5.3　How to get into Normal View

- The Slide Sorter mode displays the slides as thumbnails. This mode allows users to easily sort and organize the sequence of their slides while creating the presentation see Fig 5.4.

There are some guidelines for selecting slides in Slide Sorter view:

1) To select a single slide, just click on the slide.

2) To select a range of slides, hold down the left mouse button and drag across and over the slides like a marquee. Try to drag in a diagonal direction if users have more than one line of slides lined up.

3) To select a contiguous range of slides, click the first slide, hold the Shift key on the keyboard and then click the last slide.

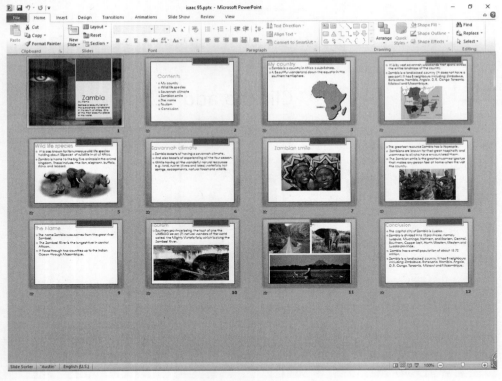

Fig 5.4　The Slide Sorter mode

4) To select multiple, noncontiguous slides, hold down the "Ctrl" key on keyboard and then click the slide user want to select.

5) To select all the slides in the presentation press "Ctrl + A" key.

6) To select all but a few slides, first select all slides, and then Ctrl click on the slides users want to deselect.

Slide Sorter view works best when users have to select multiple slides (sequential or non-sequential slides) and perform tasks such as:

1) Organize and reorder the slides by dragging them into the proper sequence.

2) Group slides into separate sections.

3) Easily copy, paste, or delete slides.

4) Duplicate slides.

5) Hide and show selected slides.

6) Control transition effects that play when the presentation advances from one slide to the next.

7) Set and adjust the slide timing of these transitions.

To change the slide order, users only need select the slide(s) people want to reorder, and drag them to a position where they want to place them.

- The Note Page mode: Besides the Comment panes below the Slide pane, users can switch to Notes Page view and enter the comments in full screen mode. So, basically its the fullscreen mode see Fig 5.5.

Fig 5.5 The Note Page mode

- The Master Views: The Slide Master, Handout Master and Notes Master note modes. They represent the main slides that stores information about the presentation, including the background, color, fonts, effects, as well as the sizes and positions of objects. The main advantage of working in a Mask mode is that users can change the universal style of each slide, notes page or document associated with their presentation see Fig 5.6.

Fig 5.6 The Master Views

1) Slide Master: in PowerPoint, users can format Master Slides for controlling their customized slides. So, what are benefits of formatting master slides? First of all, users can define the formatting for all slides at once, so the format will be consistent on all slides. Also, any changes users make to the master slides will be automatically applied to all slides in the presentation, see Fig 5.7.

If people need to open the Slide Master, they select the "View" tab→"Slide Master" from the menu ribbon. And a new Slide Master tab will automatically appear.

If people need to format Text, they can click on the master title style to select it. From the Slide Master tab, select the Fonts dropdown in the Edit Theme area.

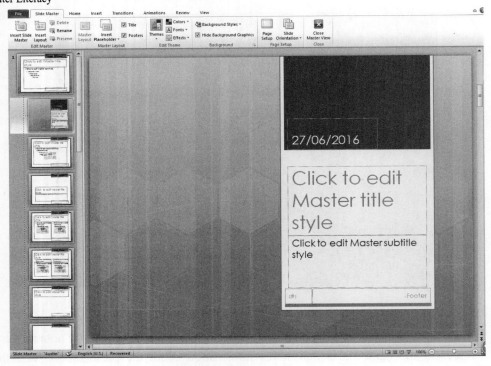

Fig 5.7　Slide Master

People can select the desired font, font style, font size, effects, and color. Click OK when people have finished selecting the font attributes for the title style and click elsewhere on the slide to deselect the title style and view the font formatting see Fig 5.8. Also, people need to repeat the steps as needed until they are satisfied with the appearance of the title style. To change the bullets for each level of text, click on the text to select it and from the Home tab, expand the Bullets and Numbering menus see Fig 5.9.

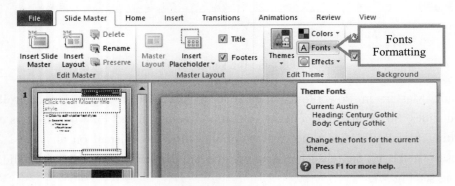

Fig 5.8　Fonts Formatting

In case of moving and resizing Text Areas, people need to click on a text area. A border appears with resize handles. To resize the text area, click and drag a resize handle while keeping the mouse clicked. A dotted line shows the new size of the area. And if people need to move the text area, move the mouse pointer over the text area. The mouse pointer becomes a four-headed arrow. Click and drag the text area to a new location while keeping the mouse clicked.

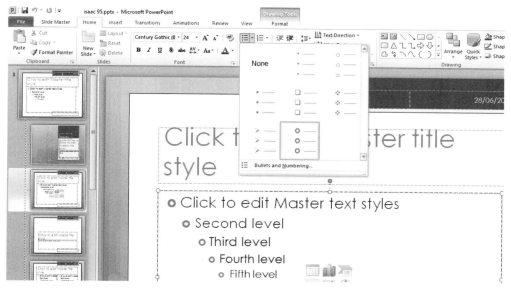

Fig 5.9   the Bullets and Numbering menus

Like Word or Excel, people also are working with Footers and Special Placeholders. The Slide Master contains placeholders for date/time, slide number, and footer. These can be moved around, resized, and reformatted. Users can also turn them on or off. To control Header or Footer, they select the "Insert" tab→"Header and Footer", which in the menu ribbon see Fig 5.10.

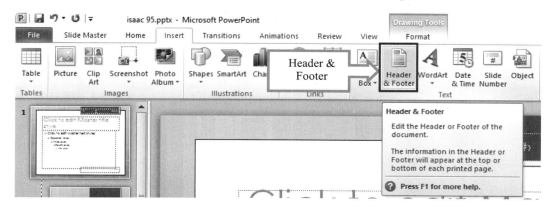

Fig 5.10   Header & Footer

When the Header and Footer box appears, click on the "Slide number" check box if users want the slides to be numbered; type text in the Footer box if users want footer text on each slide; click the "Don't show on title" slide check box if people do not want the placeholders to appear on the title slides; or by default, the date/time does not appear even though the placeholder is turned on. This is because Fixed is the default setting and the Fixed box is blank. To make the date/time appear, type the date in the Fixed box or click Update automatically to set up an automatic date/time on each slide. Click the "drop down menu" to select the desired date/time format see Fig 5.11.

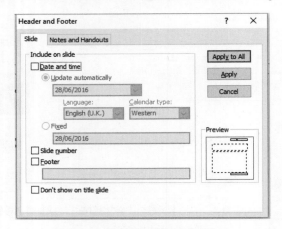

Fig 5.11 Header & Footer Dialog

When users finish setting, they click "Apply to All" to apply the footers and placeholders to all slides.

When users need to change the Background of their slides, they select the "Slide Master" tab→"Background"→"Background Styles" dropdown menu from the menu ribbon. The Background window appears see Fig 5.12.

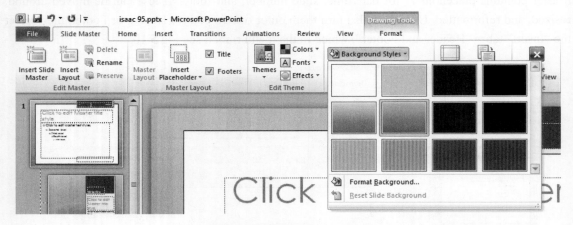

Fig 5.12 change the Background

People can select a predefined color and preview their color selections before they apply simply by mousing over the selection. At last, people click to "apply the background selection to the slide master".

After defining users' slide master formats, return to the normal view so people can edit individual slides by selecting the "Slide Master" tab→"Close Master View" from the ribbon.

2) Handout Master

Users can use the Handout Master in PowerPoint to set the appearance of all printed handouts for a presentation. To access this view, click the "Handout Master" button in the "Master Views" button group on the "View" tab in the Ribbon. This will then display the handout master for the presentation in the main window. People will also see the "Handout Master" tab appear in the Ribbon see Fig 5.13.

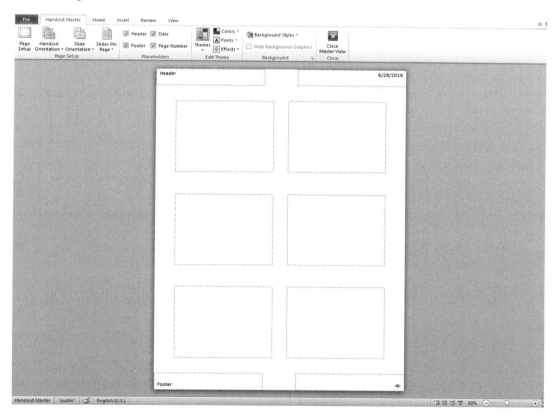

Fig 5.13　　Handout Master

People can select which handout layout to modify by selecting the desired layout from the "Slides Per Page" drop-down in the "Page Setup" button group on the "Handout Master" tab in the Ribbon. Once again, they may modify the placeholder information in the main window. They can also check or uncheck the placeholders shown in the "Placeholders" button group to add or remove those items from the layout.

3) Notes Master

Users can make changes to the notes master in PowerPoint to alter the appearance of the "Notes Page" view of presentation slides. Users can enter this view by clicking the "Notes Master" button in the "Master Views" button group ("Presentation Views" button group in 2007) on the "View" tab in the Ribbon. This will then display the notes master in PowerPoint for the current presentation within the main window. People will also see the "Notes Master" tab appear in the Ribbon see Fig 5.14.

Users can modify the content of the notes master in PowerPoint to impact the layout of the notes page view of their presentation. Users can check or uncheck the placeholders shown in the

"Placeholders" button group to add or remove those items from the layout. Once people have finished altering their notes master in PowerPoint, they can click the "Close Master View" button in the "Close" button group on the "Notes Master" tab to close the view.

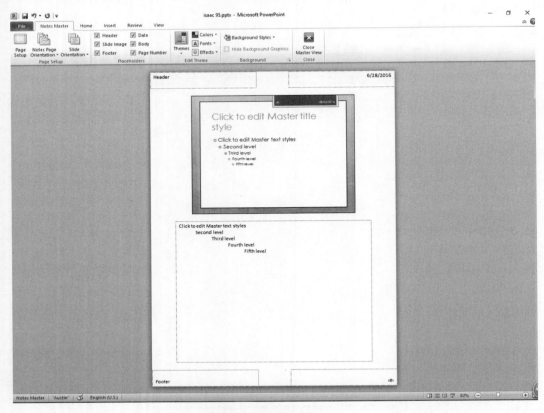

Fig 5.14　Notes Master

5.2　Creating and Formatting Slides

When a user needs to open PowerPoint, he click on the Start menu and select "All Programs" →"Microsoft Office"→"Microsoft PowerPoint 2010" and PowerPoint opens with the first slide displayed blank. Select the layout for his presentation by clicking the "Design" tab. The most commonly used design themes will appear in the center. To preview more design options, click the down arrow. The user need select a theme. Clicking it will apply the theme to all the slides in his presentation. The theme he picked will automatically create a title slide for the presentation. Click in the appropriate areas on this first slide to type title and subtitle see Fig 5.15.

5.2.1　Creating new Slide

To create a new slide, users right click in the "Navigation Pane" under any existing slide and click on the "New Slide" option, see Fig 5.16.

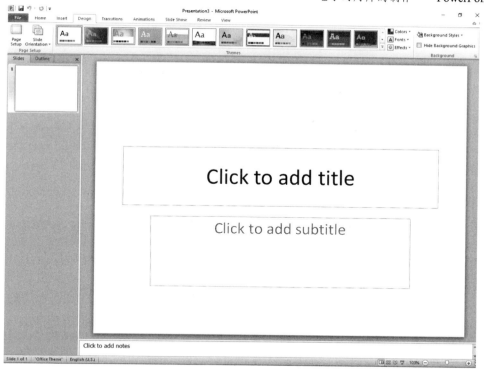

Fig 5.15　Theme Changing

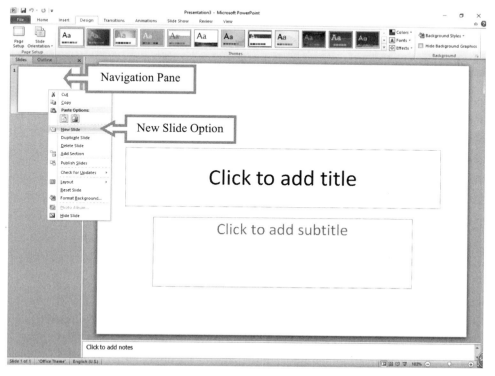

Fig 5.16　Create a new slide

The new slide is inserted. Users can now change the layout of this slide to suit the design requirements.

To change the slide layout, right click on the newly inserted slide and go to "Layout" option where users can choose from the existing layout styles available to users, see Fig 5.17.

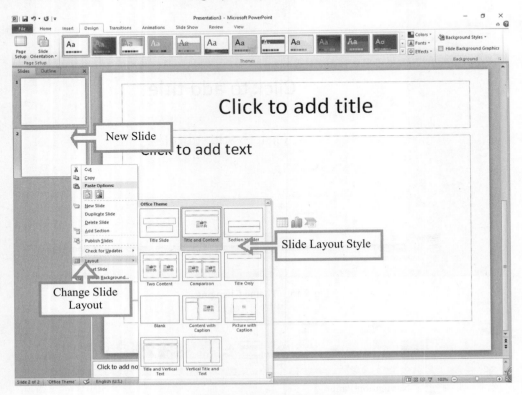

Fig 5.17 Changing Layout

Users can follow the exact same steps to insert a new slide in between existing slides or at the end on the slide list.

Usually, when users insert a new slide it will inherit the layout of its previous slide with one exception. If people are inserting a new slide after the first slide (Title slide) the subsequent slide will have Title and Content layout.

Users also need notice that if they right click in the first step without selecting any slide the menu options they get are different, although they can insert a new slide from this menu too.

5.2.2 Contents input and edit

PowerPoint allows users to add text to the slide in a well-defined manner to ensure the content is well distributed and easy to read. The procedure to add the text in a PowerPoint slide is always the same, just click in the text box and start typing. The text will follow the default formatting set for the text box, although this formatting can be changed later as required. What changes are the different kinds of content boxes that support text in a PowerPoint slide? Given below are some of the most common content blocks users will see in PowerPoint see Fig 5.18.

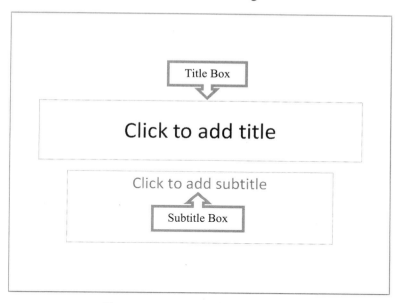

Fig 5.18   Title Slides content changing

Title Box is typically found on slides with title layout and in all the slides that have a title box in them. This box is indicated by "Click to add title" see Fig 5.19.

Subtitle Box is found only in slides with Title layout. This is indicated by "Click to add subtitle".

Fig 5.19   Title and Contents

Content Box is found in most of the slides that have a placeholder for adding content. This is indicated by "Click to add text". This box allows users to add text as well as non-text content. To add text to such a box, click anywhere on the box, except on one of the content icons in the center and start typing.

5.2.3 Choosing Layout

A typical PowerPoint presentation comprises a bunch of slides these slides usually are seeing as a blank canvas, users add their content to the slides in much the same way as they use brushes to create strokes of paint to color a canvas. However, unlike canvas, PowerPoint does not like to provide users a non-structured freedom—and this can be good in many ways. Primarily, PowerPoint categorizes each slide type into one of its prescribed layouts—examples of such layouts include(Fig 5.20):

- Title layout (comprising of placeholders to add a title and subtitle for a slide)
- Title and Content layout (comprising a slide title and a multi-purpose content placeholder)
- Title Only layout (comprising a slide title placeholder with a blank area)
- Blank layout (comprising no placeholders at all)
- And several more layouts

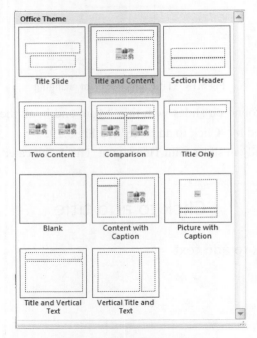

Fig 5.20   Choosing Layouts

If people need to change the layout of a slide, firstly they have to launch PowerPoint and open any existing presentation, or just use the blank presentation which is created as soon as they launch PowerPoint—such a blank presentation already includes one slid. Because each slide within a presentation has a slide layout applied to it—to ascertain which layout users active slide uses, right-click the thumbnail representing it within the Slide/Outline pane to bring up a contextual menu as shown in Fig 5.17. Within this menu, choose the Layout option—this will bring up a sub menu. Again refer to Fig 5.17, and as users can see the selected slide in this instance uses the Title Slide layout (because it's thumbnail has a yellow highlighted area). To change the layout of the active slide to another type, click on another thumbnail that represents a different layout in the same submenu.

Alternatively, select the slide whose layout users want to change and access the Home tab on the Ribbon. Within the "Slide" group click the down arrow on the right of the "Layout" button to summon the "Slide Layout" gallery. Thereafter, click any of the other slide layouts available.

5.2.4 Choosing Themes

A theme is a predefined combination of colors, fonts, and effects that can be applied to users' presentation. PowerPoint includes built-in themes that allow users to easily create professional-looking presentations without spending a lot of time formatting.

A theme is a set of colors, fonts, effects, and more that can be applied to the entire presentation to give it a consistent, professional look. People have already been using a theme, even if users didn't know it: the default Office theme, which consists of a white background, the "Calibri" font, and primarily black text. Themes can be applied or changed at any time.

Users will need to know how to apply a theme and how to switch to a different theme if users want to use this feature to create presentations. All of the themes included in PowerPoint are located in the "Themes" group on the "Design" tab. "Themes" can be applied or changed at any time.

To apply a theme, people need go to the "Design" tab and locate the "Themes" group. Each image represents a theme, like Fig 5.21.

Fig 5.21　Themes Group

Click the "drop-down" arrow to access more themes and hover over a theme to see a live preview of it in the presentation. The name of the theme will appear as people hover over it, see Fig 5.22.

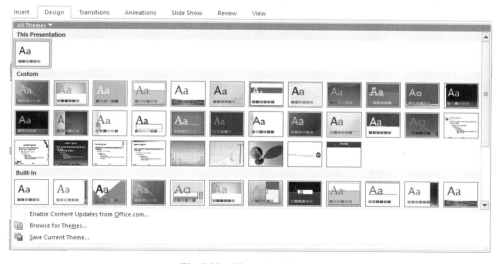

Fig 5.22　Choosing Themes

When people finishes choose a theme they can click a theme to apply it to the slides.

5.3 Formatting Slides

Users allow modifying the format or appearance of their slides. They need to add pictures, use different type of fonts, utilize tables, or show their analyzing diagrams. Therefore, to format slides can help users to bring their characters into the slides.

5.3.1 Text

One of the key elements of any good presentation is the text, hence managing the fonts in PowerPoint is vital to designing an impressive slide show. PowerPoint offers extensive font management features to cover various aspects of fonts. The font management can be accessed from the Home ribbon in the Font group, see Fig 5.23.

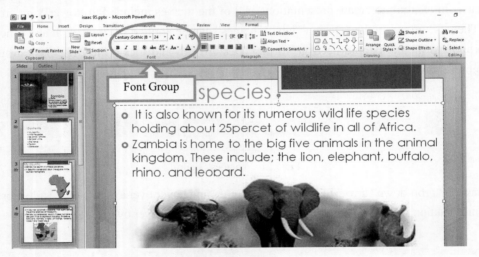

Fig 5.23　Formatting Texts

In the font group, there are the various font decoration features and their functions in PowerPoint. For examples, "B" which makes the font face bold; "*I*" means the font face italics (slanted font); "U" underlines the font face; "S" adds shadow to the font face; "abe" can strikes through the font face; will adjusts the character spacing for the font. Predefined settings are very tight, tight, normal, loose and very loose. There is a user defined spacing setting available too.

Also, users can access font management features by selecting a text box, right clicking and selecting Font, see Fig 5.24.

This opens up the Font dialog which contains all the font management features available under the font section in Home ribbon, see Fig 5.25.

Chapter 5 Microsoft PowerPoint 2010
电子幻灯片的制作——PowerPoint 2010

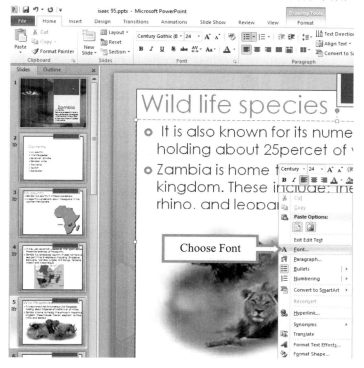

Fig 5.24   Selecting a font

Fig shows above present some functions to help users to apply different font appearance.

- Double Strikethrough: Adds two strike lines over the text.
- Superscript: Raises the text above the normal text. For example the use of "nd" in 2nd .
- Subscript: Shrinks the below the normal text. For example the '2' in H_2O the chemical formula of water.
- Small Caps: Changes the entire text to small caps.
- All Caps: Changes the entire text to capital letters.

Fig 5.25   Font dialog

- Equalize Character height: Adjusts the characters so that all are of the same height regardless of the caps setting.

The table below describes various font management features available in PowerPoint.

Table 5.1

Features	Description
Font Type	Defines the font type like Arial, Verdana, etc
Font Size	Defines the font size. Besides, there are icons to increase and decrease font size in steps in the Font group
Font Style	Defines font styles like Regular, Bold, Italics or underlined
Font Color	Specifies the font color
Font Effects	Defines effects like shadow, strikethrough, subscript, superscript, etc
Character Spacing	Specifies character spacing like loose, tight, normal, etc

5.3.2 Pictures

PowerPoint supports multiple content types including images or pictures. With regards to pictures PowerPoint classifies them into two categories:

- Picture: Images and photos that are available on users' computer or hard drive.
- Clip Art: Online picture collection that people can search from the clip art sidebar.

Although their sources are different, both these types can be added and edited in similar fashion. Go to "Images group" in the "Insert ribbon" see Fig 5.26.

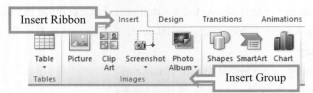

Fig 5.26　Image Ribbon

Click on "Picture" to open the "Insert Picture" dialog and add a picture to the slide.

In this dialog, users have three sections: in the left most users can browse the folders, the center section shows the subfolders and files in the selected folder and the right most section shows a preview of the selected image. Users can select the image they want and click "Open" to add the picture to the slide, see Fig 5.27.

If people need to add online pictures, click on "Clip Art" and search for keywords in the "Clip Art" sidebar see Fig 5.28.

Once users have the clipart they want to use, double click on the image to add it to the slide.

PowerPoint offers many image formatting features that can help shape the image to users' needs. The picture formatting features in PowerPoint can be accessed from the "Format" ribbon once the picture is selected. The formatting features are grouped under the "Arrange" and "Size" section in the "Format ribbon", see Fig 5.29.

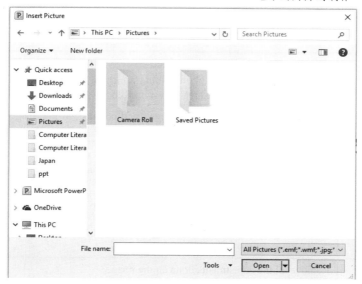

Fig 5.27　Insert Picture dialog

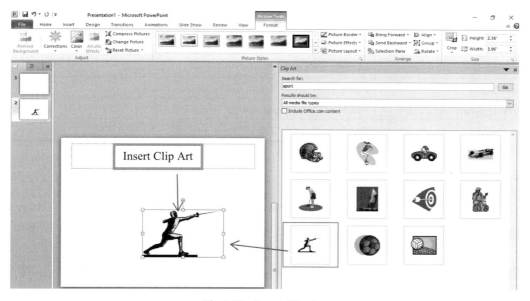

Fig 5.28　Insert Clip Art

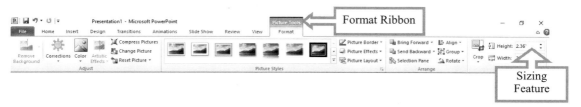

Fig 5.29　Format ribbon

The table below describes various picture adjustment features available in PowerPoint.

Table 5.2

Feature	Description
Remove Background	Automatically removes the unwanted sections in the image. This is similar to the magic tool in some of the other photo editing programs. Users can click on different regions on the image to define the area to be removed
Correction	Allows users to change the brightness and contrast on the image and also change the image sharpness
Color	Allows users to change the color on the image by changing the saturation or tone. Users can also make the image monochromatic based on different hues to match the theme of presentation
Artistic Effects	Adds artistic effects to the image like plastic wrap, glowing edges, etc.
Compress Picture	This can change the image resolution to manage the file size
Change Picture	Replace the current picture with a different one
Reset Picture	Remove all the adjustments done on the image

Picture Styles: the table below describes various picture style features available in PowerPoint.

Table 5.3

Feature	Description
Picture Border	Manage the picture border - color, weight and style
Picture Effects	Add effects to the picture like reflection, shadow, etc.
Convert to SmartArt Graphic	Transform the picture into the selected SmartArt
Quick Styles	Pre-defined styles with different picture borders and effects

5.3.3 Table

One of the most powerful data representation techniques is the use of tables. Tables allow information to be segregated making them easy to read. PowerPoint has features that let users add tables in slides and also format them to enhance their visual effects. What's more, these tables are also compatible with Microsoft Excel, so they can basically take a spreadsheet or a section of a spreadsheet and paste it into a slide as a table, see Fig 5.30.

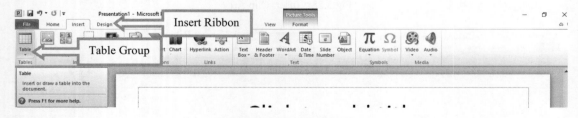

Fig 5.30　Table ribbon

People go to "Tables" group under "Insert" ribbon and click on the dropdown and select table dimension from the matrix see Fig 5.31.

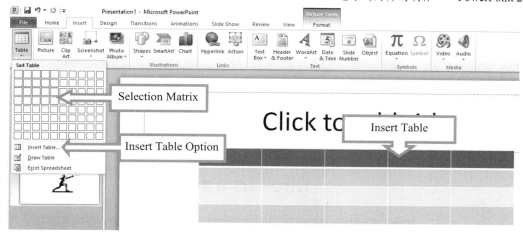

Fig 5.31   Insert Table

If a user requires more than 10 columns or 8 rows click on "Insert Table" to open the Insert Table dialog where user can specify the column and row count, see Fig 5.32.

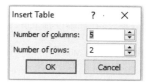

Fig 5.32   Insert Table dialog

PowerPoint table is a simple table that does not support the mathematical features of an Excel spreadsheet. If people want to carry out some calculations they can insert an Excel spreadsheet instead of a regular table by choosing it. This will insert the spreadsheet in the slide and as long as the spreadsheet is selected, the ribbon at the top would be changed to Excel ribbon instead of PowerPoint one, see Fig 5.33.

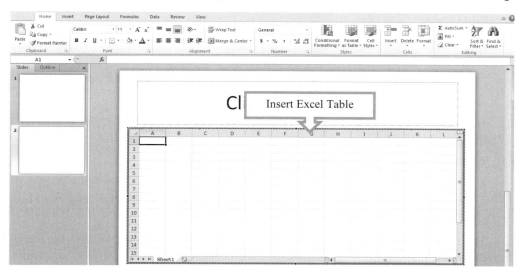

Fig 5.33   Insert a Table

The PowerPoint table formatting features have been grouped under two ribbons: Design and Format. Users can utilize the features under each ribbon. To access these ribbons people must select the table first (see Fig 5.34).

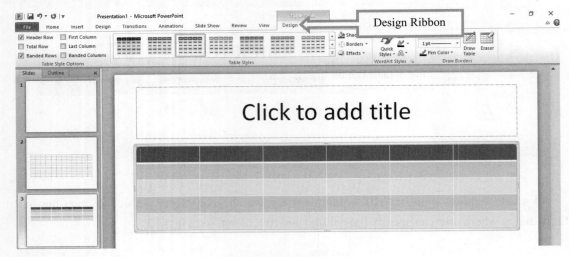

Fig 5.34 Design a table

There are various table design features in PowerPoint, which are showing below.
- Table Style Options:
  1) Header Row adds a different shade to the first row to distinguish it.
  2) Total Row adds a different shade to the last row to distinguish it.
  3) Banded Rows Shades alternate rows in the table with the same color.
  4) First Column adds a different shade to the first column to distinguish it.
  5) Last Column adds a different shade to the last row to distinguish it.
  6) Banded Columns Shades alternate columns in the table with the same color.
- Table Styles:
  1) Shading offers different shades to be added to selected table/ row/ column/ cell. Users can pick from solid shade, texture, image or gradient shading.
  2) Border offers different border options for the table. Users can edit the border color, thickness and style.
  3) Effects offer the ability to create table shadow or reflection. Users can also create bevels for individual cells.
- Word Art Styles:
  1) Text Fill allows users to change the color of the text within the table.
  2) Text Outline allows users to add an outline to the text within the table and change the outline color, weight and style.
  3) Text Effects allows users to add special effects (like reflection, shadow etc.) to the text within the table.
  4) Quick Styles contains a list of pre-defined Word Art styles that can be applied to the

selected text within the table with a single click.
- Draw Borders:
  1) Pen Style defines the style of the table border.
  2) Pen Weight defines the thickness of the table border.
  3) Pen Color defines the color of the table border.
  4) Draw Table allows users to append new rows, columns, cells to existing table, split existing rows, columns or cells and draw brand new tables.
  5) Eraser allows people to delete table borders and merge cells, rows or columns.

Table Format Features also are important to users, there are some of them shown below:
- Table Select allows users to select the entire table or the row(s) or column(s) depending on their cursor position and View Gridlines Toggle the gridline display within the table.
- Rows & Columns:
  1) Delete allows people to delete selected row(s) or column(s) or the entire table.
  2) Insert Above Inserts a row above the row where the cursor is currently. If users haven't placed the cursor within the table, it adds a new row at the top of the table
  3) Insert Below Inserts a row below the row where the cursor is currently. If people haven't placed the cursor within the table, it adds a new row at the bottom of the table.
  4) Insert Left Inserts a column to the left of the column where the cursor is currently. If people haven't placed the cursor within the table, it adds a new column to the left of the table.
  5) Insert Right Inserts a column to the right of the column where the cursor is currently. If people haven't placed the cursor within the table, it adds a new column to the right of the table.
- Merge:
  1) Merge allows users to merge cells, rows or columns. This is enabled only if they have selected more than one cell, row or column.
  2) Split Cells allows users to specify the number of rows and columns into which the current section of cell(s) need to be split.
- Cell Size:
  1) Height/Width define the height and width of the selected cell. People must realize that usually if they change these aspects for a single cell it would affect the entire row or column too.
  2) Distribute Rows equalizes the height of all the rows to fit the current table height.
  3) Distribute Columns equalizes the width of all the columns to fit the current table width.
- Alignment:
  1) Horizontal Alignment allows users to align the selected text to the left, right or center of the cell.
  2) Vertical Alignment allows users to align the selected text to the top, bottom or middle of the cell.

3) Text Direction allows user to change the direction of the selected text within the cells.

4) Cell Margins allows users to define the margins within the cell.

- Table Size:

1) Height allows users to adjust the table height—it retains the relative heights of the individual rows while changing the overall table height.

2) Width allows users to adjust the table width—it retains the relative widths of the individual columns while changing the overall table width.

3) Lock Aspect Ratio is checking this box that will ensure the ratio between the table height and width is maintained when one of these is changed.

- Arrange:

1) Bring Forward allows users to move the table up by one layer or right to the top.

2) Send Backward allows users to move the table down by one layer or right to the bottom of the slide.

3) Selection Pane toggles the Selection and Visibility sidebar.

4) Align allows users to align the entire table with reference to the slide.

5.4　Showing Effects

In PowerPoint, there are some showing effects can be utilized by users' requirements. If people need involve music, sound, cartoon, or connect with other program and website, they can use animation, hyperlink or music to enrich their presentation.

5.4.1　Animations

Benefits of adding animation and slide transitions are which provide visual interest to people's presentation and grab the audience's attention (as long as they are not overused). Users can reveal points on slides in a staggered way (i.e. one bullet displayed at a time) to keep the audience focused only on the point which they are discussing at the given time. Also, users can automate the presentation so it runs on its own.

First of all, users need to add Slide Transitions. Because slide transitions control is how the presentation moves from slide to slide.

Briefly, people do not have to add slide transitions to their presentation. If they do not add transitions, their presentation will move from slide to slide with a click of the left mouse button or by pressing Enter or the right arrow key. However, if people want the presentation to play on their own, they must set slide transitions.

How to add slide transitions, people select the slide or slides which they want to apply the transition to in either the "Slide Sorter" or "Normal" view. From the menu ribbon, people need select the "Transitions" tab.

The most commonly used animations will appear in the center. To preview more transition options, click the "down arrow", see Fig 5.35.

Fig 5.35　Transitions tab

People need select a transition from the list and click it will apply the transition to the slide. People can also select "Apply to All" to apply the same transition to all the slides. In the "Duration" option, enter the speed at which people want the transition to play. And in the "Sound" field, use the drop-down menu to select a sound to play during a slide transition, if desired, see Fig 5.36.

Fig 5.36　Timing Dialog

The first time people select a sound, a box may appear prompting them to install the Sound Effects feature. Therefore, when indicate how users want the slide transition to occur by selecting an option under the "Advance Slide" heading. They will select on mouse click if they want the transition to take place when they click the left mouse button.

If people want the transition to occur after a specified time, they select "Automatically" After. Use the "up and down arrow" keys in the blank box to specify the number of seconds which should pass before the transition takes place.

If user only need apply an animation to an object, they will choose animation to refer to how individual options on their slide move onto or off of the slide. Firstly, users select the objects within a slide that users want to apply the animations to in either the "Slide Sorter" or "Normal" view. And then they select the "Animations" tab from the menu ribbon and choose an animation, customize their play speed, and set the sound preferences. Last, users need click "Preview" to play the animations, see Fig 5.37.

Fig 5.37　Choosing Animation

PowerPoint also allows users to create "Custom Animations". People can create custom animations to specify exactly what they want to animate and how it should be done.

Users select the slide which they want to apply the animation to. Like other animation operations, from the menu bar, users select the "Animations" tab and the "Animation Pane" button. The Animation Pane appears on the right side of the screen, see Fig 5.38.

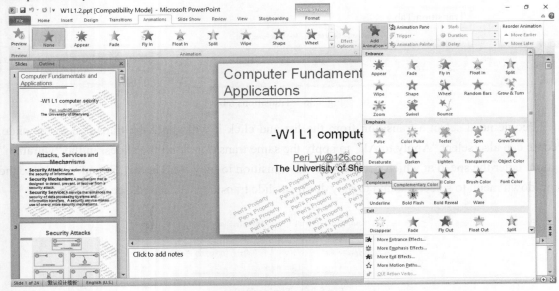

Fig 5.38   Animation Pane

In this case, if users previously selected an animation for an element on the page, the animation for it will be shown. People need to click on an element of the slide that they want to animate. For example, they could select a title, bulleted list, or graphic. Click "Add Effect", a menu appears with a list of effect categories. Entrance effects control how the element enters the slide. Emphasis effects make the element do something after it enters the slide. Exit effects control how it leaves the slide. Motion paths allow users to specify where the element travels on the slide. When users select one of the effects, details about the effect appear in the Animation Pane. Also, in the Start field, use the drop-down menu to select an event which will trigger the animation.

With Previous means the animation will occur at the same time as the previous animation on the slide or it will occur when the slide appears if there are no previous animations. And after Previous means the animation will occur after the previous animation. People allow using the lists in the Animation Pane to adjust options such as animation speed and direction.

5.4.2 Hyperlink

When people are working in Microsoft PowerPoint, they can attach material that resides outside of the PowerPoint presentation by using hyperlinks. With hyperlinks users can link part of their presentation to another slide within the same presentation, a separate PowerPoint presentation, a web site, an e-mail address or another file, like a Word document. There's a lot people can do and in this part is going to demonstrate just a few of the hyperlink options.

First, if people are going to link an element in their presentation to another slide later in the presentation. In this case, they can create a hyperlink for the link in this picture. They only need right click on the picture to highlight it and choose "Hyperlink" option or they can go up to the "Insert" tab in the top navigation and select the option to hyperlink from the links group see Fig 5.39.

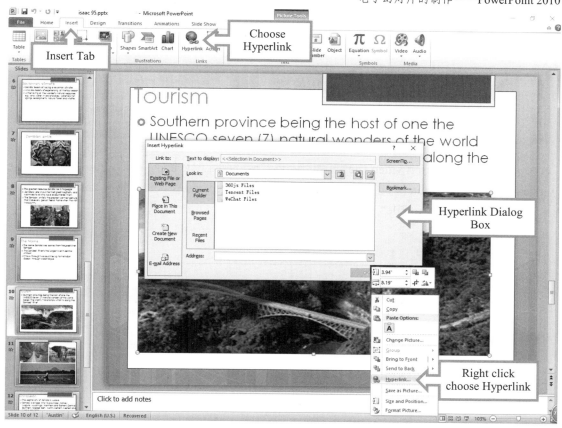

Fig 5.39　Insert Hyperlink

In the dialog box, on the left side, users can find the PowerPoint provides four different options for linking. People can link to a separate file or web page, to a different place in this document, to a new document that users create or to an e-mail address. If people need to link the picture to another slide, they can choose the second option and then select the slide they want to link to and click "OK". When users need to test, they can click up here on the slide show tab, in the top navigation and telling it to run the presentation from the current slide. They will see when they move the cursor over the picture, the arrow turns into a little hand, indicating that it's a hyperlink and if they click on the picture, they will jump to the other slide in the presentation that linked to.

Another thing people can choose is link to a file outside of the presentation. So, if people is going to hit escape to go back to normal view and then click "over to another slide" and add a hyperlink there to a word, excel or other documents.

If users ever decide they want to undo a hyperlink, they can just highlight the source again and in the insert tab when they click hyperlink; select the button that says remove hyperlink. And that's how they insert hyperlinks into a PowerPoint presentation.

5.4.3　Music

PowerPoint supports multimedia in the slides. Users can add audio or video clips to the slides which can be played during the presentation.

People allows to go to the Media group under the Insert ribbon, see Fig 5.40.

Fig 5.40   Media group

If people need to insert video file select "Video" as media type and "Video from File" to insert a video from their computers or hard drive. If people need to insert Audio they can do the same, except choose "Audio from File", see Fig 5.41.

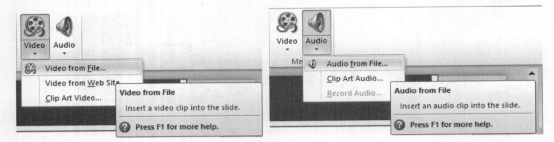

Fig 5.41   Choosing Video and Audio

In the "Insert Video/Audio" dialog, browse for a video/audio file and click "Insert" see Fig 5.42.

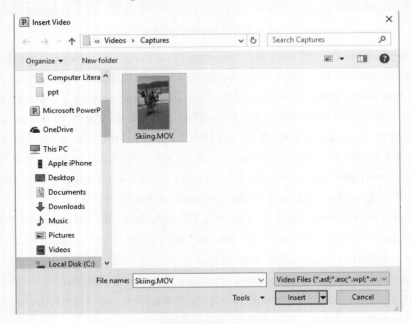

Fig 5.42   Insert Video/Audio dialog

After those operations, a video/audio file is added to the slide.

5.5 Presenting Slides

PowerPoint Slides mainly is to present people's work, idea, or progress for audience. Therefore, slides presenting is the most important part for a user. However, just showing slides is not PowerPoint can do. There are some other functions provides for users.

5.5.1 Slide show

Most PowerPoint presentations are created to be run as a slideshow. Given all the advanced features available in PowerPoint 2010, it is no surprise that there are many features related to running the slide show that have been included in this program too. Most of these features are really to help users create a good slide show without having to go through the entire presentation over and over again after every minor change. Features related to running the slide show are grouped under the "Slide Show" ribbon, see Fig 5.43.

Fig 5.43   Slide show Ribbon

Under Slide Show Ribbon, there are 3 different Groups, Start Slide Show Group, Set up Group and Monitors Group.

In Start Slide Show Group, the first function is From Beginning, which means starting slide show from beginning. From "Current Slide" will start slide show from the current slide. Broadcast Slide Show allows users to broadcast the slide shows using Microsoft's PowerPoint Broadcast Service. Custom Slide Show builds a custom slide show by picking the slides users want to run.

In Set Up Group, Set Up Slide Show can setup the slide show including browser/ full screen display, show options with or without narration/ animation, pen and laser color during the slide show and the slides to be presented during the show. Hide Slide will mark or unmark the slide as hidden so it is skipped or shown during the slide show respectively. Two very important functions are also in this group——Rehearse Timing (which allows users to rehearse the timing on each slide and the entire slide show) and Record Slide Show (which record the slide show including narration and animation). Slide Show Checkboxes Set or avoid use of narrative audio and rehearsed timings during the show. Display media controls in the slide show view.

In the Monitors Group, Resolution defines resolution in slide show view. Program will pick the monitor to display the presentation one—in case of multiple monitors. The last one, Use Presenter

View, run presentation in Presenter view rather than just slide show view.

5.5.2 Context Help

Despite getting a good grasp of the program, people may need help on different aspects from time to time. To aid in such scenarios, PowerPoint has created the context help feature. With this feature, if a user gets stuck in any dialog, he can press "F1" and PowerPoint will open the help topic related to that dialog. This is extremely beneficial as users need not spend time trying to browse through all the help topics just to get to the one they need.

The context help is based on the active window and not on the object people have selected. So if users select an image and press F1, they will get the generic help windows as their active window is still the main PowerPoint program see Fig 5.44.

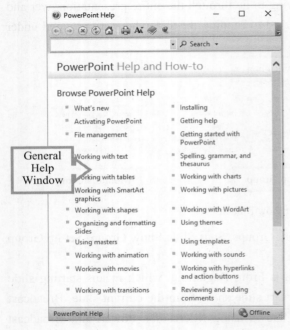
Fig 5.44   General Help Window

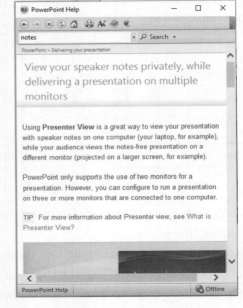
Fig 5.45   PowerPoint Help Window

If users selected any other dialog or window, PowerPoint context help will show the related help topic when they press "F1". In other words, if users continue work solely from the ribbon options, the context help would not work, but if they right click on the shapes or objects and open related editing dialogs, they can press "F1" to learn more about related functionalities, see Fig 5.45.

Chapter 6 Online Connection
网络技术

6.1 Brief introduction of Network

A network is the collection of devices that have the ability to communicate with each other. A basic understanding of networking is important for anyone managing a server. Not only is it essential for getting the services online and running smoothly, it also gives the insight to diagnose problems.

This chapter will provide a basic overview of some common networking concepts. It will also discuss basic terminology, common protocols, and the responsibilities and characteristics of the different layers of networking.

6.1.1 Network Concept

Computer networking may be considered a branch of electrical engineering, telecommunications, computer science, information technology or computer engineering, since it relies upon the theoretical and practical application of the related disciplines.

A computer network facilitates interpersonal communications allowing users to communicate efficiently and easily via various means: e-mail, instant messaging, chatting rooms, telephone, video telephone calls, and video conferencing. Providing access to information on shared storage devices is an important feature of many networks. A network allows sharing of files, data, and other types of information giving authorized users the ability to access information stored on other computers on the network. A network allows sharing of network and computing resources. Users may access and use resources provided by devices on the network, such as printing a document on a shared network printer. Distributed computing uses computing resources across a network to accomplish tasks. A computer network may be used by computer crackers to deploy computer viruses or computer worms on devices connected to the network, or to prevent these devices from accessing the network via a denial of service attack.

A computer network can be two computers connected like Fig 6.1:

Fig 6.1 2 computers connected

A computer network can also consist of, and is usually made for, more than two computers, such as Fig 6.2.

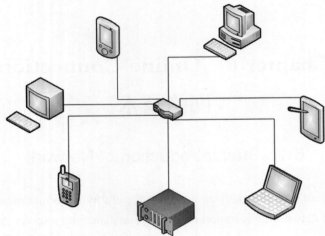

Fig 6.2　Multiple computers connected

The primary purpose of a computer network is to share resources. Therefore, some characteristics of a Computer Network show below:

- People can play a CD music from one computer while sitting on another computer.
- They may have a computer that does not have a DVD or BluRay (BD) player. In this case, people can place a movie disc (DVD or BD) on the computer that has the player, and then view the movie on a computer that lacks the player.
- People may have a computer with a CD/DVD/BD writer or a backup system but the other computer(s) does not (don't) have it. In this case, people can burn discs or make backups on a computer that has one of these but using data from a computer that does not have a disc writer or a backup system.
- People can connect a printer (or a scanner, or a fax machine) to one computer and let other computers of the network print (or scan, or fax) to that printer (or scanner, or fax machine).
- People can place a disc with pictures on one computer and let other computers access those pictures.
- People can create files and store them in one computer, then access those files from the other computer(s) connected to it.

6.1.2　History of Network

From its earliest beginnings on pages of paper and in brilliant minds, the Internet has always been an emerging technology and an emerging ideal. What follows is a selective and developing chronology of some of the most important events in the cultural and technological development of cyberspace and the Internet. Primarily intended for interested readers without a technological background, this selective chronology seeks to present a brief narrative chronology of the technological innovation of the internet and its predecessors as well as accompanying consumer and cultural developments. Due to the ongoing nature of the internet and society, this chronology is a work in progress.

Around 1960s—1970s, ARPANET commonly thought of as the predecessor to the Internet and created by the US Department of Defenses Advanced Research Projects Agency (ARPA). The first known fully operational packet-switching network, the ARPANET was designed to facilitate communication between ARPA computer terminals during the early 1960s, at a time when computers where far too expensive for widespread usage. Though conception of the idea behind ARPANET began as early as 1962, the first stable link between multiple computers through the ARPANET occurred in 1969, ten years after the first conceptual network architectural models were initiated independently by Paul Baran and Donald Davies.

In 1972, Robert Kahn exhibits the first public demonstration of the ARPANET at the International Computer Communication Conference. This public demonstration is also the first time that electronic mail (e-mail) is exhibited and is a major catalyst for increasing interest in developing network technology. The first e-mail programs called SNDMSG and READMAIL are written by Ray Tomlinson marking the beginning of one of the most widely used applications today.

While working at the Xerox Palo Alto Research Center, Robert Metcalfe develops a system which replaces radio transmission of network data with a cable that provides a larger amount of bandwidth, enabling the transfer of millions of bits of data per second in comparison with the thousands of bits per second when using a radio channel transmission from 1973 to 1975. This system is originally known as the Alto Aloha network but which was later known as Ethernet. Metcalfe would later leave Xerox to found 3Com.

In 1975, the first commercially available popular personal computer, the Altair 8800 is introduced as a kit. One year later, Apple Computer founded by Steve Jobs and Steve Wozniak.

ARPA funds the Bolt, Beranek, and Newman electronics company (intimately involved in the evolution of the ARPANET and the Internet) to use the TCP/IP protocol along with the emerging and popular Unix operating system in 1977. This is an example of the initial efforts of expanding the use of TCP/IP in the hopes of spreading a uniform standard protocol for internetworking which would allow for the emergence of the modern Internet.

Start from early 1980's, 3Com announces an Ethernet product for workstation computers followed by a version for personal computers in 1982. The 3Com Ethernet networking system allows users to build more affordable local area networks (LANs) and has come to be the standard way of building personal networks. In the same year, IBM releases its own Personal Computer (PC) which will become the industry standard. Sun Microsystems launches the Sun I workstation.

ARPANET is split into two separate entities: MILNET to serve the needs of national defense and the military community and ARPANET which is primarily utilized by academics and researchers. This is the first step toward the commercialization of the Internet. The unofficial birth of the Internet occurs when both ARPANET and Defense Data Networks begin to use TCP/IP protocol. After an increase in the number of users and networks connected to the Internet, computer scientists begin to consider new methods of naming and addressing users and networks. The Domain Name System, first defined by Paul Mockapetris, Jon Postel, and Craig Partridge is created to divide Internet names between host names (x.com) and user names (user@x.com). This method of naming made the maintenance of host information more manageable. ARPA created six domains consisting initially of: edu (educational), gov (government), mil (military), com (commercial), org (other organizations), and net (network resources).

In 1984, Apple Macintosh is launched and the first domain is registered in 1985, see Fig 6.3.

Fig 6.3　Apple Macintosh

Performance Systems International begins offering TCP/IP network services to businesses in 1989. Later, some scientists begin to create the first actual incarnation of the World Wide Web. Berners-Lee and his colleagues developed a shared format for hypertext documents which was named hypertext markup language or HTML. In addition to HTML, Berners-Lee and others created uniform resource locator (URL) as a standard address format that could specify the computer being targeted and the type of information being requested. URL and HTML significantly increased the possibility of interaction between users and networks across the Internet. URL also made different Internet services such as Usenet news accessible to all users employing the system.

By 1995, the bulk of US internet traffic is routed through interconnected network service providers and Microsoft Windows 95 is launched. 1995 proved to be an eventful year in the formation of contemporary Internet culture because it also sees the official launch of the online bookstore Amazon.com, the Internet search engine Yahoo, online auction site Ebay, the Internet Explorer web browser by Microsoft, and the creation by Sun Microsystems of the Java programming language which allows for the programming of animation on websites giving rise to a new level of Internet interactivity.

6.1.3　Network Classification

Computer network can be classified on the different criteria, such as Scale, Connection Method, Functional Relationship (Network Architectures), or Network Topology.

Nowadays, connection method is helping people to divide common network in a clear way. Computer networks can be classified according to the hardware and software technology that is used to interconnect the individual devices in the network, such as Optical fiber, Ethernet, wireless LAN,

HomePNA, Power line communication or Ethernet uses physical wiring to connect devices. Frequently deployed devices include hubs, switches, bridges, or routers. Wireless LAN technology is designed to connect devices without wiring. These devices use radio waves or infrared signals as a transmission medium. ITU-G.hn technology uses existing home wiring (Coaxial cable, Phone lines and Power lines) to create a high-speed (up to 1Gigabit/s) local area network.

- Wired technologies

Twisted pair wire is the most widely used medium for telecommunication. Twisted-pair wires are ordinary telephone wires which consist of two insulated copper wires twisted into pairs and are used for both voice and data transmission. The use of two wires twisted together helps to reduce crosstalk and electromagnetic induction. The transmission speed ranges from 2 million bits per second to 100 million bits per second.

Coaxial cable is widely used for cable television systems, office buildings, and other worksites for local area networks. The cables consist of copper or aluminum wire wrapped with insulating layer typically of a flexible material with a high dielectric constant, all of which are surrounded by a conductive layer. The layers of insulation help minimize interference and distortion. Transmission speed range from 200 million to more than 500 million bits per second.

Optical fiber cable consists of one or more filaments of glass fiber wrapped in protective layers. It transmits light which can travel over extended distances. Fiber-optic cables are not affected by electromagnetic radiation. Transmission speed may reach trillions of bits per second. The transmission speed of fiber optics is hundreds of times faster than for coaxial cables and thousands of times faster than a twisted-pair wire.

- Wireless technologies

Terrestrial microwave: Terrestrial microwaves use Earth-based transmitter and receiver. The equipment looks similar to satellite dishes. Terrestrial microwaves use low-gigahertz range, which limits all communications to line-of-sight. Path between relay stations spaced approx, 30 miles apart. Microwave antennas are usually placed on top of buildings, towers, hills, and mountain peaks.

Communications satellites: The satellites use microwave radio as their telecommunications medium which are not deflected by the Earth's atmosphere. The satellites are stationed in space, typically 22,000 miles (for geosynchronous satellites) above the equator. These Earth orbiting systems are capable of receiving and relaying voice, data, and TV signals.

Cellular and PCS systems: Use several radio communications technologies. The systems are divided to different geographic areas. Each area has a low-power transmitter or radio relay antenna device to relay calls from one area to the next area.

There is another division criterion which can show the basic structure of network—By Scale. The hardware and transformation protocol are different when they are using in the different scale of network. Therefore, computer networks may be classified according to the scale, in Local Area Network (LAN), Metropolitan Area Network (MAN), and Wide Area Network (WAN).

- Local Area Network (LAN)

A Local Area Network is a computer network covering a small Networks geographical area, like a home, office, or groups of buildings e.g. a school Network. For example, a library will have a wired or wireless LAN Network for users to interconnect local networking devices e.g., printers and servers to connect to the internet. The defining characteristics of LANs Network, in contrast to Wide Area Networks (WANs), includes their much higher data-transfer rates, smaller geographic range, and lack of need for leased telecommunication lines. Although switched Ethernet is now the most common protocol for Networks. Current Ethernet or other IEEE 802.3 LAN technologies operate at speeds up to 10 Gbit/s. IEEE has projects investigating the standardization of 100 Gbit/s, and possibly 40 Gbit/s. Smaller Networks generally consist of a one or more switches linked to each other——often with one connected to a router, cable modem, or DSL modem for Internet access. LANs Network may have connections with other LANs Network via leased lines, leased services (Fig 6.4).

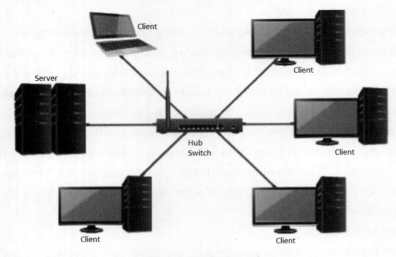

Fig 6.4　LANs Network

　　LAN is a group of computers located in the same room, on the same floor or in the same building that are connected to form a single network as to share resources such as disk drives, printers, data, CPU, fax/modem, application. etc.

　　LAN is generally limited to specific geographical area less than 2 K.M., supporting high speed networks. A wide variety of LANs have been built and installed, but a few types have more recently become dominant. The most widely used LAN system is the Ethernet system based on the bus topology.

　　Intermediate nodes (i.e., repeaters, bridges and switches) allow to be connected together to from larger LANs. A LAN may also be connected to another LAN or to WANs and MANs using a "router" device.

　　There are essentially five components of a LAN. Firstly, network devices such as Workstations, printers, file servers which are normally accessed by all other computers. Secondly, network communication device, such as hubs, routers, switches etc. Those can be used for network connectivity. Network Interface Cards (NICs) for each network device required to access the network. It is the interface between the machine and the physical network. Cable is used as a physical transmission medium. Last, network operating system, software applications required to control the use of network operation and administration.

The characteristics of LAN are it connects computers in a single building, block or campus, i.e. they work in a restricted geographical area. LANs' are private networks, not subject to tariffs or other regulatory controls. LAN's operate at relatively high speed when compared to the typical WAN.

Advantages of LAN are it allows sharing of expensive resources such as Laser printers, software and mass storage devices among a number of computers. LAN allows for high-speed exchange of essential information. It contributes to increased productivity. A LAN installation should be studied closely in the context of its proposed contribution to the long range interest of the organization.

Disadvantages of LAN are the financial cost of LAN is still high in comparison with many other alternatives. It requires memory space in each of the computers used on the network. This reduces the memory space available for the user's programs. Some type of security system must be implemented if it is important to protect confidential data. Some control on the part of the user is lost. People may have to share a printer with other users. They may face a situation like, for example, the entire network suddenly locking up because one user has made a mistake.

- Metropolitan Area Network (MAN)

Metropolitan area networks, or MANs, are large computer network usually spanning a city. They typically use wireless infrastructure or Optical fiber connections to link their sites (Fig 6.5).

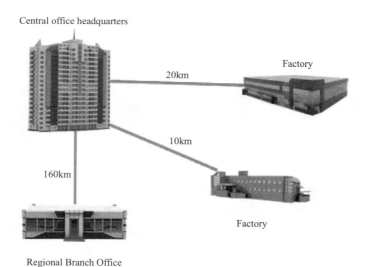

Fig 6.5 Metropolitan Area Network

A MAN is optimized for a larger geographical area than a LAN, ranging from several blocks of buildings to entire cities. MANs can also depend on communications channels of moderate-to-high data rates. A MAN might be owned and operated by a single organization, but it usually will be used by many individuals and organizations. MANs might also be owned and operated as public utilities or privately owned. They will often provide means for internetworking of local networks. Metropolitan area networks can span up to 50km, devices used are modem and wire/cable.

A Metropolitan Area Network is a large computer network that spans a metropolitan area or campus. Its geographic scope falls between a WAN and LAN. MANs provide Internet connectivity for LANs in a metropolitan region, and connect them to wider area networks like the Internet. The network size falls intermediate between LAN and WAN. A MAN typically covers an area of between 5 and 50 km diameter. Many MANs cover an area the size of a city, although in some cases MANs may be as small as a group of buildings or as large as the North of Scotland. A MAN often acts as a high speed network to allow sharing of regional resources. It is also frequently used to provide a shared connection to other networks using a link to a WAN.

The characteristics of MAN are it generally covers towns and cities (50 kms). It is developed in 1980s. Communication medium used for MAN are optical fibers, cables etc. Data rates adequate for distributed computing applications.

- Wide Area Network (WAN)

Wide Area Network is a network system connecting cities, countries or continents, a network that uses routers and public communications links. The largest and most well-known example of a WAN is the Internet, see Fig 6.6.

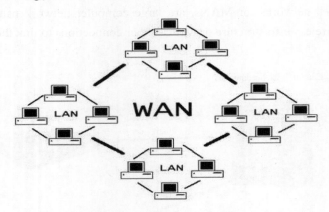

Fig 6.6 Wide Area Network

WANs are used to connect LANs and other types of networks together, so that users and computers in one location can communicate with users and computers in other locations. Many WANs are built for one particular organization and are private. Others, built by Internet service providers, provide connections from an organization's LAN to the Internet. WANs are often built using leased lines. At each end of the leased line, a router connects to the LAN on one side and a hub within the WAN on the other. Leased lines can be very expensive. Instead of using leased lines, WANs can also be built around public network or Internet.

The Characteristics of WAN are it generally covers large distances (states, countries, continents). Communication medium used are satellite, public telephone networks which are connected by routers. Routers forward packets from one to another a route from the sender to the receiver.

In each city, LAN, WAN, and MAN are consisting together to accomplish the whole network for normal people using. The structure diagram shows below (Fig 6.7).

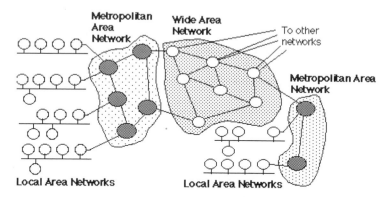

Fig 6.7   LAN, WAN, and MAN connection

6.1.4   Network Protocol

A network protocol defines rules and conventions for communication between network devices. Protocols for computer networking all generally use packet switching techniques to send and receive messages in the form of packets.

Network protocols include mechanisms for devices to identify and make connections with each other, as well as formatting rules that specify how data is packaged into messages sent and received. Some protocols also support message acknowledgement and data compression designed for reliable and/or high-performance network communication. Hundreds of different computer network protocols have been developed each designed for specific purposes and environments.

OSI protocols are a family of standards for information exchange. These were developed and designed by the International Organization of Standardization (ISO). In 1977 the ISO model was introduced, which consisted of seven different layers. This model has been criticized because of its technicality and limited features.

Each layer of the ISO model has its own protocols and functions. The OSI protocol stack was later adapted into the TCP/IP stack. In some networks, protocols are still popular using only the data link and network layers of the OSI model.

The OSI protocol stack works on a hierarchical form, from the hardware physical layer to the software application layer. There are a total of seven layers. Data and information are received by each layer from an upper layer. After the required processing, this layer then passes the information on to the next lower layer. A header is added to the forwarded message for the convenience of the next layer. Each header consists of information such as source and destination addresses, protocol used, sequence number and other flow-control related data (see Fig 6.8).

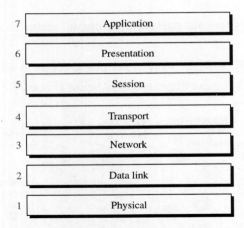

Fig 6.8 the seven layers of the OSI Model

- Layer 1, the Physical Layer: This layer deals with the hardware of networks such as cabling. The major protocols used by this layer include Bluetooth, PON, OTN, DSL, IEEE.802.11, IEEE.802.3, L431 and TIA 449.
- Layer 2, the Data Link Layer: This layer receives data from the physical layer and compiles it into a transform form called framing or frame. The protocols are used by the Data Link Layer include: ARP, CSLIP, HDLC, IEEE.802.3, PPP, X-25, SLIP, ATM, SDLS and PLIP.
- Layer 3, the Network Layer: This is the most important layer of the OSI model, which performs real time processing and transfers data from nodes to nodes. Routers and switches are the devices used for this layer. The network layer assists the following protocols: Internet Protocol (IPv4), Internet Protocol (IPv6), IPX, AppleTalk, ICMP, IPSec and IGMP.
- Layer 4, the Transport Layer: The transport layer works on two determined communication modes: Connection oriented and connectionless. This layer transmits data from source to destination node. It uses the most important protocols of OSI protocol family, which are: Transmission Control Protocol (TCP), UDP, SPX, DCCP and SCTP.
- Layer 5, the Session Layer: The session layer creates a session between the source and the destination nodes and terminates sessions on completion of the communication process. The protocols used are: PPTP, SAP, L2TP and NetBIOS.
- Layer 6, the Presentation Layer: The functions of encryption and decryption are defined on this layer. It converts data formats into a format readable by the application layer. The following are the presentation layer protocols: XDR, TLS, SSL and MIME.
- Layer 7, the Application Layer: This layer works at the user end to interact with user applications. QoS (quality of service), file transfer and e-mail are the major popular services of the application layer. This layer uses following protocols: HTTP, SMTP, DHCP, FTP, Telnet, SNMP and SMPP

OSI layered framework for the design of network systems that allows communication across all types of computer systems. Vertical and horizontal communication between the layers using interfaces. (Defines what information and services should the layer provide to the layer above it.)

When an user sends/transmits a data to the another user, then here is the diagram (Fig 6.9) that shows how data is sent as well as received:

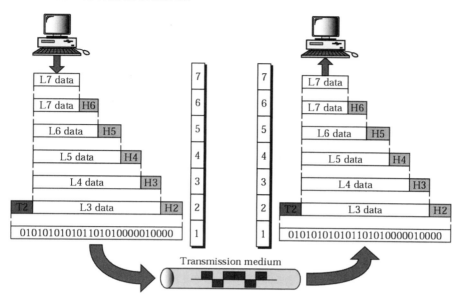

Fig 6.9   Data Transmission

As the diagram showing, the data need to be encapsulated from each layer. Firstly, PDU conception is providing each protocol on the different layer which has its own format. And, headers are added while a packet is going down the stack at each layer. At the last, trailers are usually added on the second layer.

6.2   Local Area Network (LAN)

A Local Area Network (LAN) is a group of computers and associated devices that share a common communications line or wireless link to a server. Typically, a LAN encompasses computers and peripherals connected to a server within a distinct geographic area such as an office or a commercial establishment. Computers and other mobile devices use a LAN connection to share resources such as a printer or network storage.

A Local Area Network may serve as few as two or three users (for example, in a small office network) or several hundred users in a larger office. LAN networking comprises cables, switches, routers and other components that let users connect to internal servers, websites and other LANs via wide area networks.

Ethernet and Wi-Fi are the two primary ways to enable LAN connections. Ethernet is a specification that enables computers to communicate with each other. Wi-Fi uses radio waves to connect computers to the LAN. Other LAN technologies, including Token Ring, Fiber Distributed

Data Interface and ARCNET, have lost favor as Ethernet and Wi-Fi speeds have increased. The rise of virtualization has fueled the development of virtual LANs, which allows network administrators to logically group network nodes and partition their networks without the need for major infrastructure changes (see Fig 6.10).

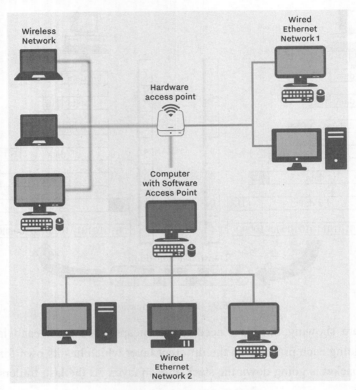

Fig 6.10　Local Area Network

Typically, a suite of application programs can be kept on the LAN server. Users who need an application frequently can download it once and then run it from their local device. Users can order printing and other services as needed through applications run on the LAN server. A user can share files with others stored on the LAN server; read and write access is maintained by a network administrator. A LAN server may also be used as a web server if safeguards are taken to secure internal applications and data from outside access.

6.2.1　Structure of LAN

The structure of LAN is following the structure of network, also been called topology of network. Network topology is the arrangement of the various elements (links, nodes, etc.) of a computer network. Essentially, it is the topological structure of a network and may be depicted physically or logically. Physical topology is the placement of the various components of a network, including device location and cable installation, while logical topology illustrates how data flows within a network, regardless of its physical design. Distances between nodes, physical interconnections, transmission rates, or signal types may differ between two networks, yet their topologies may be identical.

There are two basic categories of network topologies: physical topologies and logical topologies.

The cabling layout used to link devices is the physical topology of the network. This refers to the layout of cabling, the locations of nodes, and the interconnections between the nodes and the cabling. The physical topology of a network is determined by the capabilities of the network access devices and media, the level of control or fault tolerance desired, and the cost associated with cabling or telecommunications circuits.

The logical topology in contrast, is the way that the signals act on the network media, or the way that the data passes through the network from one device to the next without regard to the physical interconnection of the devices. A network's logical topology is not necessarily the same as its physical topology. For example, the original twisted pair Ethernet using repeater hubs was a logical bus topology with a physical star topology layout. Token Ring is a logical ring topology, but is wired as a physical star from the Media Access Unit.

The study of network topology recognizes several basic topologies: bus, star, ring or circular, mesh, tree, hybrid, or daisy chain.

- Bus: In local area networks where bus topology is used, each node is connected to a single cable, by the help of interface connectors. This central cable is the backbone of the network and is known as the bus (thus the name). A signal from the source travels in both directions to all machines connected on the bus cable until it finds the intended recipient. If the machine address does not match the intended address for the data, the machine ignores the data. Alternatively, if the data matches the machine address, the data is accepted. Because the bus topology consists of only one wire, it is rather inexpensive to implement when compared to other topologies. However, the low cost of implementing the technology is offset by the high cost of managing the network. Additionally, because only one cable is utilized, it can be the single point of failure (see Fig 6.11).

Fig 6.11　Bus Topology

- Star network topology: In LAN with a star topology, each network host is connected to a central hub with a point-to-point connection. So it can be said that every computer is indirectly connected to every other node with the help of the hub. In Star topology, every node (computer workstation or any other peripheral) is connected to a central node called hub, router or switch. The switch is the server and the peripherals are the clients. The network does not necessarily have to resemble a star to be classified as a star network, but all of the nodes on the network must be connected to one central device. All traffic that traverses the network passes through the central hub. The hub acts as a signal repeater. The star topology is considered the easiest topology to design and implement. An advantage of the star topology is the simplicity of adding additional nodes. The primary disadvantage of the star topology is that the hub represents a single point of failure. (Fig 6.12)

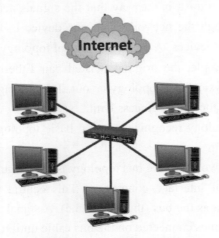

Fig 6.12   Star Topology

- Ring network topology: A network topology is set up in a circular fashion in such a way that they make a closed loop. This way data travels around the ring in one direction and each device on the ring acts as a repeater to keep the signal strong as it travels. Each device incorporates a receiver for the incoming signal and a transmitter to send the data on to the next device in the ring. The network is dependent on the ability of the signal to travel around the ring. When a device sends data, it must travel through each device on the ring until it reaches its destination. Every node is a critical link. In a ring topology, there is no server computer present; all nodes work as a server and repeat the signal. The disadvantage of this topology is that if one node stops working, the entire network is affected or stops working (Fig 6.13).
- Mesh networking: The value of fully meshed networks is proportional to the exponent of the number of subscribers, assuming that communicating groups of any two endpoints, up to and including all the endpoints, is approximated by Reed's Law (Fig 6.14).
- Fully connected mesh topology: In a fully connected network, all nodes are interconnected. (In graph theory this is called a complete graph.) The simplest fully connected network is a two-node network. A fully connected network doesn't need to use packet switching or

broadcasting. However, since the number of connections grows quadratically with the number of nodes (Fig 6.15):

$$c = n/(n-1)$$

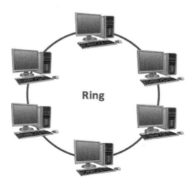

Fig 6.13   Ring Topology

Fig 6.14   Mesh Topology

This makes it impractical for large networks.

Fig 6.15   Fully connected mesh Topology

In generally, there is no one topology better than the others. Based on the running environment and users' requirements, all types of topology can be used separately or mixed.

6.2.2 LAN Components

Local Area Networks connects computers together to exchange data. Apart from the computers, and other devices like printers and faxes, a LAN has to have six essential components to function.

- Network Adapter: A computer needs a network adapter to connect to a network (see Fig 6.16). It converts computer data into electronic signals. It listens for silence on the network cable and applies the data to it when it has an opportunity. The network access element of its job is called Media Access Control, or MAC. The physical address of every computer on a network is called its MAC address. The MAC address is the network adapter's serial number. Most computers are shipped with the network adapter integrated into the motherboard. However, early PCs didn't include this function and computer owners had to buy it separately and fit it into an expansion slot on the motherboard. These were called "network cards" because they were sold on a separate card. Although network adapters are now integrated, the name network card is still used. The wireless equivalent is called a Wireless Network Interface Controller.

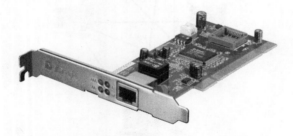

Fig 6.16 Network Adapter

- Network Medium: Wired networks need cable (Fig 6.17). The most common form of cable used in networks is called the "Unshielded Twisted Pair". In PC shops, it is generally just referred to as "network cable" or "Ethernet cable". Ethernet is the most widely implemented set of standards for the physical properties of networks. UTP is so closely identified with Ethernet that it is often given that name. Other cable types used for networks are twin-axial, Shielded Twisted Pair and single-mode and multi-mode fiber optic cable. Wireless networks don't need cable; they send data on radio waves generated by the WNIC.
- Cable Connectors: In wired networks, the most common form of connector is the RJ45 (Fig 6.18). Every computer with networking capabilities has an RJ45 port. This is sometimes called a "network port" or an "Ethernet port". The RJ45 plug looks like a slightly larger telephone plug and connects the Unshielded Twisted Pair or the Shielded Twisted Pair cable.

Chapter 6 Online Connection
网络技术

Fig 6.17   Network Medium   Fig 6.18   Cable Connectors

- Power Supply: Both wired and wireless networks need a power supply. A wireless network uses the current to generate radio waves. A cabled network sends data interpreted as an electronic pulse.
- Hub/Switch/Router: In wired networks, one computer cannot connect to many others without some form of splitter (Fig 6.19). A hub is little more than a splitter. It repeats any signals coming into one of its ports out onto all its other ports. A cable leads from each port to one computer. A switch is a more sophisticated version of a hub. It only sends the signal on to the computer with the address written in the arriving message. Routers are much more complicated and are able to forward messages all over the world. Larger networks sometimes use routers for their LAN traffic. The wireless networking device is called a "Wireless Router." (see Fig 6.20)

Fig 6.19   Hub

Fig 6.20   Wireless Router

- Network Software: Software on a communicating computer packages data into segments and puts that data into a structure called a "packet." The source and destination addresses of the packet are written into the header of the packet. The receiving computer needs to interpret these packets back into meaningful data and deliver it to the appropriate application.

6.3 Internet acknowledges

The Internet, sometimes called simply "the Net", is a worldwide system of computer networks— a network of networks in which users at any one computer can, if they have permission, get information from any other computer (and sometimes talk directly to users at other computers). It was conceived by the Advanced Research Projects Agency (ARPA) of the U.S. government in 1969 and was first known as the ARPANET. The original aim was to create a network that would allow users of a research computer at one university to "talk to" research computers at other universities. A side benefit of ARPANET's design was that, because messages could be routed or rerouted in more than one direction, the network could continue to function even if parts of it were destroyed in the event of a military attack or other disaster.

6.3.1 Introduction of the Internet

Today, the Internet is a public, cooperative and self-sustaining facility accessible to hundreds of millions of people worldwide. Physically, the Internet uses a portion of the total resources of the currently existing public telecommunication networks. Technically, what distinguishes the Internet is its use of a set of protocols called TCP/IP (for Transmission Control Protocol/Internet Protocol). Two recent adaptations of Internet technology, the intranet and the extranet, also make use of the TCP/IP protocol.

For most Internet users, electronic mail (e-mail) practically replaced the postal service for short written transactions. People communicate over the Internet in a number of other ways including Internet Relay Chat (IRC), Internet telephony, instant messaging, video chat or social media.

The most widely used part of the Internet is the World Wide Web (often abbreviated "WWW" or called "the Web"). Its outstanding feature is hypertext, a method of instant cross-referencing. In most Web sites, certain words or phrases appear in text of a different color than the rest; often this text is also underlined. When people select one of these words or phrases, they will be transferred to the site or page that is relevant to this word or phrase. Sometimes there are buttons, images, or portions of images that are "clickable". If they move the pointer over a spot on a Web site and the pointer changes into a hand, this indicates that users can click and be transferred to another site.

Using the Web, people have access to billions of pages of information. Web browsing is done with a Web browser, the most popular of which are Chrome, Firefox and Internet Explorer. The appearance of a particular Web site may vary slightly depending on the browser they use. Also, later versions of a particular browser are able to render more "bells and whistles" such as animation, virtual reality, sound, and music files, than earlier versions.

The Internet has continued to grow and evolve over the years of its existence. IPv6, for example, was designed to anticipate enormous future expansion in the number of available IP addresses. In a related development, the Internet of Things (IoT) is the burgeoning environment in which almost any entity or object can be provided with a unique identifier and the ability to transfer data automatically over the Internet.

6.3.2 Client/Server format

A client-server network is a central computer, also known as a server, which hosts data and other forms of resources. Clients such as laptops and desktop computers contact the server and request to use data or share its other resources with it (Fig 6.21).

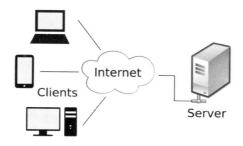

Fig 6.21 A client-server network

A client-server network is designed for end-users, called clients, to access resources such as files, songs, video collections, or some other service from a central computer called a server. A server's sole purpose is to do what its name implies - serve its clients.

In general, a service is an abstraction of computer resources and a client does not have to be concerned with how the server performs while fulfilling the request and delivering the response. The client only has to understand the response based on the well-known application protocol, i.e. the content and the formatting of the data for the requested service.

Clients and servers exchange messages in a request–response messaging pattern: The client sends a request, and the server returns a response. This exchange of messages is an example of inter-process communication. To communicate, the computers must have a common language, and they must follow rules so that both the client and the server know what to expect. The language and rules of communication are defined in a communications protocol. All client-server protocols operate in the application layer. The application-layer protocol defines the basic patterns of the dialogue. To formalize the data exchange even further, the server may implement an API (such as a web service). The API is an abstraction layer for such resources as databases and custom software. By restricting communication to a specific content format, it facilitates parsing. By abstracting access, it facilitates cross-platform data exchange.

A server may receive requests from many different clients in a very short period of time.

Because the computer can perform a limited number of tasks at any moment, it relies on a scheduling system to prioritize incoming requests from clients in order to accommodate them all in turn. To prevent abuse and maximize uptime, the server's software limits how a client can use the server's resources. Even so, a server is not immune from abuse. A denial of service attack exploits a server's obligation to process requests by bombarding it with requests incessantly. This inhibits the server's ability to respond to legitimate requests.

The biggest advantage to using this setup is central management of the server. Only one server is used to host the resources that all the clients request and use. This is especially good for server administrators, because they only have to be in one place and can solve all the problems in one place, as well. Having to manually update several hundred servers would take much more time. One centrally managed server is the key to ease of management, and it is cost effective, too.

Another advantage of using one physical server is that the configuration is simple to set up and takes less time to troubleshoot. For instance, if there were a site with multiple servers providing redundant services, and it was having issues, it could take an extreme amount of work to effectively troubleshoot why services are being hindered. In a single server role, all troubleshooting takes place at one physical server, so it takes much less time.

6.3.3 TCP/IP Protocol

Transmission Control Protocol/Internet Protocol (TCP/IP) is the language which a computer uses to access the Internet. It consists of a suite of protocols designed to establish a network of networks to provide a host with access to the Internet. TCP/IP is responsible for full-fledged data connectivity and transmitting the data end-to-end by providing other functions, including addressing, mapping and acknowledgment. TCP/IP contains four layers, which differ slightly from the OSI model.

The technology is so common that people would rarely refer to somebody use the full name. In other words, in common usage the acronym is now the term itself. Nearly all computers today support TCP/IP. TCP/IP is not a single networking protocol, it is a suite of protocols named after the two most important protocols or layers within it TCP and IP.

As with any form of communication, two things are needed: a message to transmit and the means to reliably transmit the message. The TCP layer handles the message part. The message is broken down into smaller units, called packets, which are then transmitted over the network. The packets are received by the corresponding TCP layer in the receiver and reassembled into the original message.

The IP layer is primarily concerned with the transmission portion. This is done by means of a unique IP address assigned to each and every active recipient on the network.

TCP/IP is considered a stateless protocol suite because each client connection is newly made without regard to whether a previous connection had been established.

6.3.4 IP address

The IP address definition is based on Internet Protocol Version 4. An IPv4 address consists of four sets of numbers from 0 to 255, separated by three dots. For example, the IP address of TechTerms.com is 67.43.14.98. This number is used to identify the TechTerms website on the Internet. The total number of IPv4 addresses ranges from 000.000.000.000 to 255.255.255.255. Because $256 = 2^8$, there are $2^8 \times 4$ or 4,294,967,296 possible IP addresses. While this may seem like a large number, it is no longer enough to cover all the devices connected to the Internet around the world. Therefore, many devices now use IPv6 addresses.

See Internet Protocol Version 6 (IPv6) for a description of the newer 128-bit IP address. Note that the system of IP address classes described here, while forming the basis for IP address assignment, is generally bypassed today by use of Classless Inter-Domain Routing (CIDR) addressing. The IPv6 address format is much different than the IPv6 format. It contains eight sets of four hexadecimal digits and uses colons to separate each block. An example of an IPv6 address is: 2602:0445:0000:0000:a93e:5ca7:81e2:5f9d. There are 3.4×10^{38} or 340 undecillion) possible IPv6 addresses.

In the most widely installed level of the Internet Protocol (IP) today, an IP address is a 32-bit number that identifies each sender or receiver of information that is sent in packets across the Internet. When people request an HTML page or send e-mail, the Internet Protocol part of TCP/IP includes people's IP address in the message (actually, in each of the packets if more than one is required) and sends it to the IP address that is obtained by looking up the domain name in the Uniform Resource Locator people requested or in the e-mail address they are sending a note to. At the other end, the recipient can see the IP address of the Web page requestor or the E-mail sender and can respond by sending another message using the IP address it received.

An IP address has two parts: the identifier of a particular network on the Internet and an identifier of the particular device (which can be a server or a workstation) within that network. On the Internet itself—that is, between the routers that move packets from one point to another along the route—only the network part of the address is looked at.

- The Network Part of the IP Address: the Internet is really the interconnection of many individual networks (it's sometimes referred to as an internetwork). So the Internet Protocol (IP) is basically the set of rules for one network communicating with any other (or occasionally, for broadcast messages, all other networks). Each network must know its own address on the Internet and that of any other networks with which it communicates. To be part of the Internet, an organization needs an Internet network number, which it can request from the Network Information Center (NIC). This unique network number is included in any packet sent out of the network onto the Internet.
- The Local or Host Part of the IP Address: In addition to the network address or number, information is needed about which specific machine or host in a network is sending or receiving a message. So the IP address needs both the unique network number and a host number (which is unique within the network). (The host number is sometimes called a local or machine address.)

Part of the local address can identify a sub-network or subnet address, which makes it easier for a network that is divided into several physical sub-networks (for examples, several different local area networks) to handle many devices.

Since networks vary in size, there are four different address formats or classes to consider when applying to NIC for a network number:
- Class A addresses are for large networks with many devices.
- Class B addresses are for medium-sized networks.
- Class C addresses are for small networks (fewer than 256 devices).
- Class D addresses are multicast addresses.

The first few bits of each IP address indicate which of the address class formats it is using. The address structures look, like Fig 6.22:

Class A

0	Network (7 bits)	Local address (24 bits)

Class B

10	Network (14 bits)	Local address (16 bits)

Class C

110	Network (21 bits)	Local address (8 bits)

Class D

1110	Multicast address (28 bits)

Fig 6.22   An Example of web browser

The IP address is usually expressed as four decimal numbers, each representing eight bits, separated by periods. This is sometimes known as the dot address and, more technically, as dotted quad notation. For Class A IP addresses, the numbers would represent "network.local.local.local"; for a Class C IP address, they would represent "network.network.network.local". The number version of the IP address can (and usually is) represented by a name or series of names called the domain name.

The Internet's explosive growth makes it likely that, without some new architecture, the number of possible network addresses using the scheme above would soon be used up (at least, for Class C network addresses). However, a new IP version, IPv6, expands the size of the IP address to 128 bits, which will accommodate a large growth in the number of network addresses. For hosts still using IPv4, the use of subnets in the host or local part of the IP address will help reduce new applications for network numbers. In addition, most sites on today's mostly IPv4 Internet have gotten around the Class C network address limitation by using the Classless Inter-Domain Routing (CIDR) scheme for address notation.

Also, there are relationships between the IP Address and the Physical Address. The machine or physical address used within an organization's local area networks may be different than the Internet's IP address. The most typical example is the 48-bit Ethernet address. TCP/IP includes a facility called the Address Resolution Protocol (ARP) that lets the administrator create a table that maps IP addresses to physical addresses. The table is known as the ARP cache.

In some area, static IP address is commonly used. By contrast, dynamic IP Address is also used under certain situation. The discussion above assumes that IP addresses are assigned on a static basis. In fact, many IP addresses are assigned dynamically from a pool. Many corporate networks and online services economize on the number of IP addresses they use by sharing a pool of IP addresses among a large number of users.

6.4 Internet Application

Billions of applications are applying in the Internet, which help users understand or use the Internet in better place. Internet services are provided by application software which is an application on one computer communicates across a network with an application program running on another computer. The Internet applications span a wide range, such as e-mail, file transfer, web browsing, voice telephone calls (VoIP), distributed databases, audio/video teleconferencing, and etc. Each application offers a specific service to the user using a specific user interface. But all applications can communicate over a single, shared network.

6.4.1 World Wide Web

The Web, or World Wide Web, is basically a system of Internet servers that support specially formatted documents. The documents are formatted in a markup language called HTML (Hypertext Markup Language) that supports links to other documents, as well as graphics, audio, and video files.

A broader definition comes from the organization that Web inventor Tim Berners-Lee helped found, he said: "The World Wide Web (W3C) is the universe of network-accessible information, an embodiment of human knowledge."

The web browser is another Internet application of critical importance. It was developed and then standardized in the early. The web browser was developed in a highly commercialized environment dominated by such corporations as Microsoft and Netscape, and heavily influenced by the World Wide Web Consortium (W3C). While Microsoft and Netscape have played the most obvious parts in the development of the web browser, particularly from the public perspective, the highly influential role of the W3C may be the most significant in the long term.

Founded in 1994 by Tim Berners-Lee, the original architect of the web, the goal of the W3C has been to develop interoperable technologies that lead the web to its full potential as a forum for communication, collaboration, and commerce. What the W3C has been able to do successfully is to develop and promote the adoption of new, open standards for web-based documents. These standards have been designed to make web documents more expressive (Cascading Stylesheets), to provide standardized labeling so that users have a more explicit sense of the content of documents (Platform for Internet Content Selection, or PICS), and to create the basis for more interactive designs (the Extensible Markup Language, or XML). Looking ahead, a principal goal of the W3C is to develop capabilities that are in accordance with Berners-Lee's belief that the web should be a highly collaborative information space.

Microsoft and Netscape dominate the market for web browsers, with Microsoft's Internet Explorer holding about three quarters of the market, and Netscape holding all but a small fraction of the balance. During the first few years of web growth, the competition between Microsoft and Netscape for the browser market was fierce, and both companies invested heavily in the development of their respective browsers. Changes in business conditions toward the end of the 1990s and growing interest in new models of networked information exchange caused each company to focus less intensely on the development of web browsers, resulting in a marked slowing of their development and an increasing disparity between the standards being developed by W3C and the support offered by Internet Explorer or Netscape Navigator.

Now, the future of the web browser may be short-lived, as standards developers and programmers elaborate the basis for network-aware applications that eliminate the need for the all purpose browser. It is expected that as protocols such as XML and the Simple Object Access Protocol (SOAP) grow more sophisticated in design and functionality, an end user's interactions with the web will be framed largely by desktop applications called in the services of specific types of documents called from remote sources.

The open source model has important implications for the future development of web browsers. Because open source versions of Netscape have been developed on a modular basis, and because the source code is available with few constraints on its use, new or improved services can be added quickly and with relative ease. In addition, open source development has accelerated efforts to integrate web browsers and file managers. These efforts, which are aimed at reducing functional distinctions between local and network-accessible resources, may be viewed as an important element in the development of the "seamless" information space that Berners-Lee envisions for the future of the web.

6.4.2 E-mail

E-mail (electronic mail) is the exchange of computer−stored messages by telecommunication. (Some publications spell it e-mail; we prefer the currently more established spelling of e-mail.) E-mail messages are usually encoded in ASCII text. However, users can also send non-text files, such as graphic images and sound files, as attachments sent in binary streams. E-mail was one of the first uses of the Internet and is still the most popular use. A large percentage of the total traffic over the Internet is e-mail. E-mail can also be exchanged between online service provider users and in networks other than the Internet, both public and private.

E-mail can be distributed to lists of people as well as to individuals. A shared distribution list can be managed by using an e-mail reflector. Some mailing lists allow users to subscribe by sending a request to the mailing list administrator. A mailing list that is administered automatically is called a list server.

E-mail is one of the protocols included with the Transport Control Protocol/Internet Protocol (TCP/IP) suite of protocols. A popular protocol for sending e-mail is Simple Mail Transfer Protocol and a popular protocol for receiving it is POP3. Both Netscape and Microsoft include an e-mail utility with their Web browsers.

Fig 6.23   An Example of E-mail

Whether judged by volume, popularity, or impact, e-mail has been and continues to be the principal Internet application. This is despite the fact that the underlying technologies have not been altered significantly since the early 1980s. In recent years, the continuing rapid growth in the use and volume of e-mail has been fueled by two factors. The first is the increasing numbers of Internet Service Providers (ISPs) offering this service, and secondly, because the number of physical devices capable of supporting e-mail has grown to include highly portable devices such as personal digital assistants (PDAs) and cellular telephones.

The volume of e-mail also continues to increase because there are more users, and because users now have the ability to attach documents of various types to e-mail messages. While this has long been possible, the formulation of Multipurpose Internet Mail Extensions (MIME) and its adoption by software developers has made it much easier to send and receive attachments, including word-processed documents, spreadsheets, and graphics. The result is that the volume of traffic generated by e-mail, as measured in terms of the number of data packets moving across the network, has increased dramatically in recent years, contributing significantly to network congestion.

E-mail has become an important part of personal communications for hundreds of millions of people, many of whom have replaced it for letters or telephone calls. In business, e-mail has become an important advertising medium, particularly in instances where the demand for products and services is time sensitive. For example, tickets for an upcoming sporting event are marketed by sending fans an e-mail message with information about availability and prices of the tickets. In addition, e-mail serves, less obviously, as the basis for some of the more important collaborative applications that have been developed, most notably Lotus Notes.

In the near future, voice-driven applications will play a much larger role on the Internet, and e-mail is sure to be one of the areas in which voice-driven applications will emerge most rapidly. E-mail and voice mail will be integrated, and in the process it seems likely that new models for Internet-based messaging will emerge.

Synchronous communication, in the form of the highly popular "instant messaging", may be a precursor of the messaging models of the near future. Currently epitomized by AOL Instant Messenger and Microsoft's Windows Messenger, instant messaging applications generally allow users to share various types of files (including images, sounds, URLs), stream content, and use the Internet as a medium for telephony, as well as exchanging messages with other users in real time and participating in online chat rooms.

How can a user use an e-mail? Firstly, he needs to apply an e-mail account with Internet connection. He should find which e-mail provider is satisfied by his requirements. Different e-mail providers contain different performance such as bigger store space, faster transmission speed, or more stable condition than the others. Users do not worry about which one is the best and they will hardly find which one is the prefect one. What they have to do is to find the e-mail provider which suitable for them, for example, Gmail (Fig 6.24).

Fig 6.24 Creating new account Page of Gmail

When users find the correct e-mail provider, they can create their own account. They need to write down some of their personnel information to create a new account (see Fig 6.25).

Chapter 6 Online Connection
网络技术

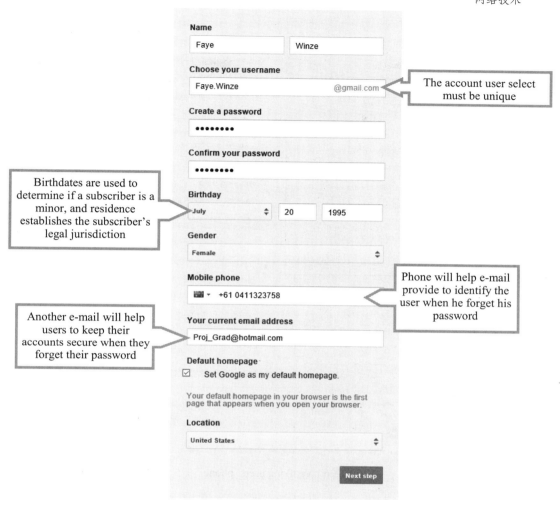

Fig 6.25 Legal issues Dialog

When all the information fill in the blanks and is accepted by the web, users can click "Next step" to continue their accounts creation. Usually, the e-mail providers will pump out a dialog to prevent their legal issues.

When users pull the vertical scroll all the way down to the bottom, "I AGREE" button will be activated. If people click the bottom, it means users agree with the legal requirements.

After verifying users' phone, people can use their own Gmail account. Faye.Winze@gmail.com will be the account's name, and other people will connect him with it (Fig 6.26 and Fig 6.27).

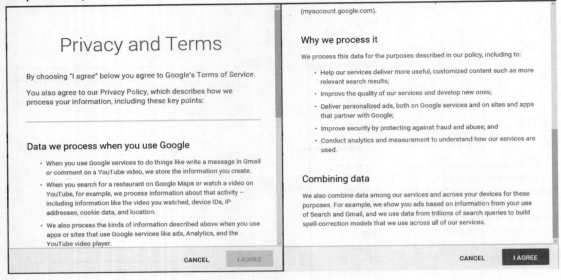

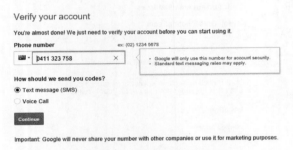

Fig 6.26   verifying users' phone

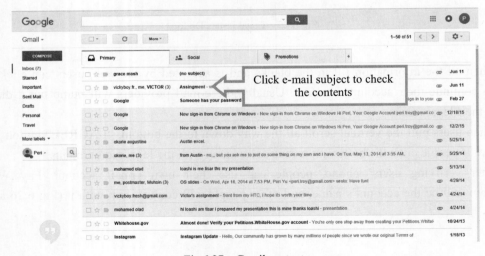

Fig 6.27   Gmail contents

When users open their e-mail account, they will find some e-mails are already in their box.

They can check e-mail or reply e-mail (like Fig 6.28).

Fig 6.28 Gmail Index Page

If users need to write a new e-mail to someone, they can just click the red compose button in any page to do it.

Even though, e-mail providers give users plenty of e-mail address to use, some of them still love to use some combined e-mail account, for example, the QQ users automatically have a QQ e-mail account when they apply their QQ account. The QQ account is QQ number@qq.com.

6.4.3 File download

The Internet is a wonderful resource for obtaining files of any imaginable type of data. When users utilize Web browsers, they may follow these steps to how a file is downloaded from the Internet.

First of all, to initiate a download from a website, users must first click on the link. A link is usually denoted by an icon or different colored text.

Next, the browser should ask what users would like to do with the file. Unless they want to run the program immediately, it is recommended people save the file to a location on their computer that people are going to remember, like the desktop.

If the file users have downloaded is executable, double click the "icon" to start the setup process. However, the majority of the time, a downloaded file is compressed and requires another program to extract the file's contents before setup can begin. Fortunately, this function is built into later versions of Windows. Once the file has been extracted, double click the "setup icon" to install. see Fig 6.29 and Fig 6.30.

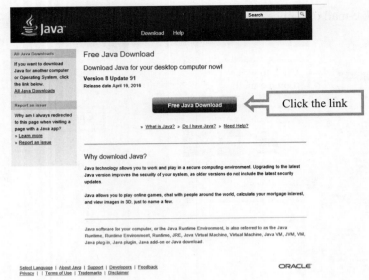

Fig 6.29　Java Download Webpage

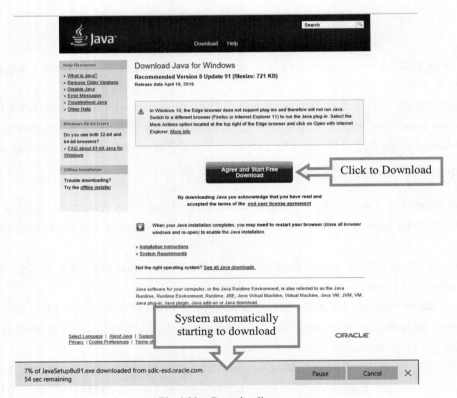

Fig 6.30　Downloading

6.4.4　Search on Internet

As talked before, Internet has billions of useful knowledge or software. However, how people can find them is a beginner's big concerned. Searching useful information is powerful technics for every user.

Firstly, users need to select a search engine. At the top of any page on the computer, type the phrase "search engines" into the Search Bar to attain access to several different internet sites that specifically aid in searching. Common search engines are Bing, Google, Yahoo, or Baidu. Fig 6.31 is Google search engine.

Fig 6.31   Google Mainpage

After typing in whatever users want to search, press the Enter key on the keyboard of their computer.

Choose a few of the most specific or relevant keywords or phrases to describe the topic. Utilize synonyms. Type their choice of words into the Search Bar offered by their chosen search engine. Generally, capitalization and punctuation are not needed and search engines usually disregard minor words such as "the, and, to, etc."

Assess the results. Search through the list of web pages to pinpoint information (Fig 6.32).

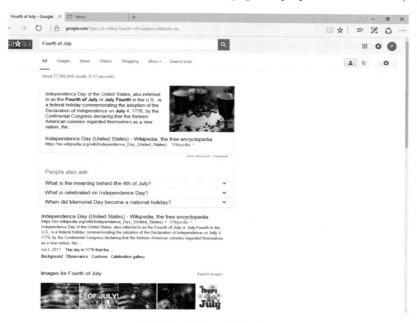

Fig 6.32   Google Search

People can repeat above steps as necessary when they need to choose a different search engine or choose new search words that are more or less specific.

Use the Advanced Search found on most sites. For example, if users want to go to Advanced Search on Google, go here (see Fig 6.33).

Fig 6.33   Google Advanced Search

It is not correct to assume that people's subject is more or less equally visible in all search engines so it is not important which one to use. Recent engines sort pages also by rank that is assigned is a complex, ever changing, usually secret way and is different for every search company. While engines will likely be "consistent" for highly popular web sites, less popular web sites may be ranked very differently and it may make sense to try multiple engines.

6.5   Network Security

Network security has become more important to personal computer users, organizations, and the military. With the advent of the internet, security became a major concern and the history of security allows a better understanding of the emergence of security technology. The internet structure itself allowed for many security threats to occur. The architecture of the Internet, when modified can reduce the possible attacks that can be sent across the network. Knowing the attack methods, allows for the appropriate security to emerge. Many businesses secure themselves from the internet by means of firewalls and encryption mechanisms. The businesses create an "intranet" to remain connected to the internet but secured from possible threats. The entire field of network security is vast and in an evolutionary stage. The range of study encompasses a brief history dating back to internet's beginnings and the current development in network security. In order to understand the research being performed today, background knowledge of the internet, its vulnerabilities, attack methods through the internet, and security technology is important and therefore they are reviewed.

There are 3 key terms in network security, which are security attacks, services and mechanisms.
- Security Attack: Any action that compromises the security of information.
- Security Mechanism: A mechanism that is designed to detect, prevent, or recover from a security attack.

- Security Service: A service that enhances the security of data processing systems and information transfers. A security service makes use of one or more security mechanisms.

All 3 parts are tightly combined together and none of them can be ignored in any fields.

6.5.1 Security Attack

When considering network security, it must be emphasized that the whole network is secure. Network security does not only concern the security in the computers at each end of the communication chain. When transmitting data the communication channel should not be vulnerable to attack. A possible hacker could target the communication channel, obtain the data, and decrypt it and reinsert a false message. Securing the network is just as important as securing the computers and encrypting the message. When developing a secure network, the following need to be considered:

- Access: Authorized users are provided the means to communicate to and from a particular network.
- Confidentiality: Information in the network remains private.
- Authentication: Ensure the users of the network are who they say they are.
- Integrity: Ensure the message has not been modified in transit.
- Non-repudiation: Ensure the user does not refute that he used the network.

Base on those considerable requirements, there are 4 different types of security attacks just mapping them.

- Interruption: This is an attack on availability to interrupt the transformation data flow.
- Interception: This is an attack on confidentiality. It allows data flow to normally transport with copy or eavesdropping.
- Modification: This is an attack on integrity. After an attacker has read data, the next logical step is to alter it. An attacker can modify the data in the packet without the knowledge of the sender or receiver. Even if people do not require confidentiality for all communications, they do not want any of their messages to be modified in transit.
- Fabrication: This is an attack on authenticity. Most networks and operating systems use the IP address of a computer to identify a valid entity. In certain cases, it is possible for an IP address to be falsely assumed—identity spoofing. An attacker might also use special programs to construct IP packets that appear to originate from valid addresses inside the corporate intranet.

After gaining access to the network with a valid IP address, the attacker can modify, reroute, or delete people's data.

Therefore, how a system can protect users avoid such attacks? Some methods of defense should be introduced:

- Encryption: encryption is the process of encoding messages or information in such a way that only authorized parties can read it. The purpose of encryption is to ensure that only somebody who is authorized to access data (e.g. a text message or a file); It will be able to read it, using the decryption key. Somebody who is not authorized can be excluded, because he or she does not have the required key, without which it is impossible to read the encrypted information. (Fig 6.34)

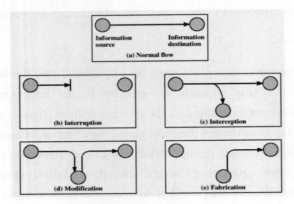

Fig 6.34 Security Attacks

- Software Controls: Access limitations in a data base, in operating system protect each user from other users.
- Hardware Controls: People can use smartcard or other identity devices to proof themselves.
- Policies: the system can provide serious rules to increase the security level, for example frequent changes of passwords.
- Physical Controls: Improper installation, selecting wrong components, incomplete devices, lack of knowledge, unsecure or less secure network components can cause physical threat to the critical network resources. Physical threats are divided in two types; accidentally and intentionally. With proper planning users can minimize accidental damage.

6.5.2 Internet Virus

In computers, a virus is a program or programming code that replicates by being copied or initiating its copying to another program, computer boot sector or document. Viruses can be transmitted as attachments to an e-mail note or in a downloaded file, or be present on a diskette or CD. The immediate source of the e-mail note, downloaded file, or diskette people have received is usually unaware that it contains a virus. Some viruses wreak their effect as soon as their code is executed; other viruses lie dormant until circumstances cause their code to be executed by the computer. Some viruses are benign or playful in intent and effect ("Happy Birthday!") and some can be quite harmful, erasing data or causing hard disk to require reformatting. A virus that replicates itself by resending itself as an e-mail attachment or as part of a network message is known as a worm.

Generally, there are three main classes of viruses:
- File infectors. Some file infector viruses attach themselves to program files, usually selected .COM or .EXE files. Some can infect any program for which execution is requested, including .SYS, .OVL, .PRG, and .MNU files. When the program is loaded, the

virus is loaded as well. Other file infector viruses arrive as wholly-contained programs or scripts sent as an attachment to an e-mail note.
- System or boot-record infectors. These viruses infect executable code found in certain system areas on a disk. They attach to the DOS boot sector on diskettes or the Master Boot Record on hard disks. A typical scenario (familiar to the author) is to receive a diskette from an innocent source that contains a boot disk virus. When operating system is running, files on the diskette can be read without triggering the boot disk virus. However, if people leave the diskette in the drive, and then turn the computer off or reload the operating system, the computer will look first in the A drive, find the diskette with its boot disk virus, load it, and make it temporarily impossible to use hard disk. (Allow several days for recovery.) This is why users should make sure they have a bootable floppy.
- Macro viruses. These are among the most common viruses, and they tend to do the least damage. Macro viruses infect users' Microsoft Word application and typically insert unwanted words or phrases.

Some types of computer viruses are shown below:
- A logical bomb is a destructive program that performs an activity when a certain action has occurred.
- A worm is also a destructive program that fills a computer system with self-replicating information, clogging the system so that its operations are slowed down or stopped.
- A boot sector virus infects boot sector of computers. During system boot, boot sector virus is loaded into main memory and destroys data stored in hard disk.
- A macro virus is associated with application software like word and excel. When opening the infected document, macro virus is loaded into main memory and destroys the data stored in hard disk.
- Trojan horse is a destructive program. It usually pretends as computer games or application software. If executed, computer system will be damaged.

The best protection against a virus is to know the origin of each program or file people load into their computer or open from their e-mail program. Since this is difficult, they can buy anti-virus software that can screen e-mail attachments and also check all of files periodically and remove any viruses that are found. From time to time, people may get an e-mail message warning of a new virus. Unless the warning is from a source they recognize, chances are good that the warning is a virus hoax.

6.5.3 Anti-Virus Software

Antivirus (or anti-virus) software is used to safeguard a computer from malware, including viruses, computer worms, and Trojan horses. Antivirus software may also remove or prevent spyware and adware, along with other forms of malicious programs. Free antivirus software generally only searches their computer using signature-based detection which involves looking for patterns of data that are known to be related to already-identified malware. Paid antivirus software will usually also include heuristics to catch new, or zero-day threats, by either using genetic signatures to identify new variants of existing virus code or by running the file in a virtual environment (also called a sandbox), and watching what it does to see if it has malicious intent.

Virus designers, however, usually test their malicious code against the major antivirus types of malware, specifically ransomware, use polymorphic code to make it difficult to be detected by antivirus software. Besides using antivirus software to keep their computer safe and running smoothly, it is also always a good idea to be proactive: make sure the web browser is updated to the latest version, use a firewall, only download programs from websites people trust and always surf the web using a standard user account, rather than administrator one.

In antivirus products market, there are plenty of different antivirus software, for example, McAfee, Norton, Kaspersky, AVIRA and etc.

6.6 New in Internet

Since Internet appears in human's lives, it develops itself every second. From early stage people's communication to big data analysis currently, Internet always is one of the most important modern tools for everyone. Recently, Internet based technologies affects people's normal life, there are 2 interesting technologies will be introduced in this part.

6.6.1 Cloud

In telecommunications, a cloud is the unpredictable part of any network through which data passes between two end points. Possibly the term originated from the clouds used in blackboard drawings or more formal illustrations to describe the non-specifiable or uninteresting part of a network. Clouds exist because between any two points in a packet-switched network, the physical path on which a packet travels can vary from one packet to the next and, in a circuit switched network, the specific circuit that is set up can vary from one connection to the next.

After cloud has been involved into people's lives, cloud computing became a top term for a lot of individuals and business.

In basic terms, cloud computing is the phrase used to describe different scenarios in which computing resource is delivered as a service over a network connection (usually, this is the internet). Cloud computing is therefore a type of computing that relies on sharing a pool of physical and/or virtual resources, rather than deploying local or personal hardware and software. It is somewhat synonymous with the term 'utility computing' as users are able to tap into a supply of computing resource rather than manage the equipment needed to generate it themselves; much in the same way as a consumer tapping into the national electricity supply, instead of running their own generator (shown as Fig 3.35).

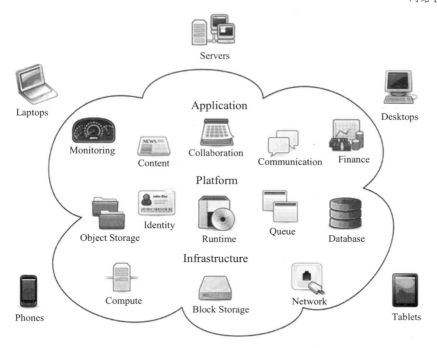

Fig 6.35   Cloud Computing

One of the key characteristics of cloud computing is the flexibility that it offers and one of the ways that flexibility is offered is through scalability. This refers to the ability of a system to adapt and scale to changes in workload. Cloud technology allows for the automatic provision and deprovision of resource as and when it is necessary, thus ensuring that the level of resource available is as closely matched to current demand as possible. This is a defining characteristic that differentiates it from other computing models where resource is delivered in blocks (e.g., individual servers, downloaded software applications), usually with fixed capacities and upfront costs. With cloud computing, the end user usually pays only for the resource they use and so avoids the inefficiencies and expense of any unused capacity.

However, the advantages of cloud computing are not limited to flexibility. Enterprise can also benefit (in varying degrees) from the economies of scale created by setting up services en masse with the same computing environments, and the reliability of physically hosting services across multiple servers where individual system failures do not affect the continuity of the service.

There is also great choice in the level of security and management required in cloud deployments, with an option to suit almost any business:
- A public cloud, for example, is a cloud in which services and infrastructure are hosted off-site by a cloud provider, shared across their client base and accessed by these clients via public networks such as the internet. Public clouds offer great economies of scale and redundancy but are more vulnerable than private cloud setups due their high levels of accessibility.

- Private clouds on the other hand use pooled services and infrastructure stored and maintained on a private network—whether physical or virtual—accessible for only one client. The obvious benefits to this are greater levels of security and control. Cost benefits must be sacrificed to some extent though, as the enterprise in question will have to purchase/rent and maintain all the necessary software and hardware.
- The final cloud option is a hybrid cloud and this, as the name suggests, combines both public and private cloud elements. A hybrid cloud allows a company to maximize their efficiencies; by utilizing the public cloud for non-sensitive operations while using a private setup for sensitive or mission critical operations, companies can ensure that their computing setup is ideal without paying any more than is necessary.

6.6.2 Big Data

Big data is a buzzword, or catch-phrase, meaning a massive volume of both structured and unstructured data that is so large it is difficult to process using traditional database and software techniques. In most enterprise scenarios the volume of data is too big or it moves too fast or it exceeds current processing capacity.

Despite these problems, big data has the potential to help companies improve operations and make faster, more intelligent decisions. This data, when captured, formatted, manipulated, stored, and analyzed can help a company to gain useful insight to increase revenues, get or retain customers, and improve operations.

Therefore, a lot of people are holding the question: is Big Data Volume or a Technology?

While the term may seem to reference the volume of data, that isn't always the case. The term big data, especially when used by vendors, may refer to the technology (which includes tools and processes) that an organization requires handling the large amounts of data and storage facilities. The term big data is believed to have originated with Web search companies who needed to query very large distributed aggregations of loosely-structured data.

An example of big data might be petabytes (1,024 terabytes) or exabytes (1,024 petabytes) of data consisting of billions to trillions of records of millions of people—all from different sources (e.g. Web, sales, customer contact center, social media, mobile data and so on). The data is typically loosely structured data that is often incomplete and inaccessible.

Also many minds believe big data can control types of Business Datasets. When dealing with larger datasets, organizations face difficulties in being able to create, manipulate, and manage big data. Big data is particularly a problem in business analytics because standard tools and procedures are not designed to search and analyze massive datasets.

Big data can be characterized by 3Vs: the extreme volume of data, the wide variety of types of data and the velocity at which the data must be must processed. Although big data doesn't refer to any specific quantity, the term is often used when speaking about petabytes and exabytes of data, much of which cannot be integrated easily.

Because big data takes too much time and costs too much money to load into a traditional relational database for analysis, new approaches to storing and analyzing data have emerged that rely less on data schema and data quality. Instead, raw data with extended metadata is aggregated in a data lake and machine learning and artificial intelligence (AI) programs use complex algorithms to look for repeatable patterns.

Big data analytics is often associated with cloud computing because the analysis of large data sets in real-time requires a platform like Hadoop to store large data sets across a distributed cluster and MapReduce to coordinate, combine and process data from multiple sources.

Although the demand for big data analytics is high, there is currently a shortage of data scientists and other analysts who have experience working with big data in a distributed, open source environment. In the enterprise, vendors have responded to this shortage by creating Hadoop appliances to help companies take advantage of the semi-structured and unstructured data they own.

Big data can be contrasted with small data, another evolving term that's often used to describe data whose volume and format can be easily used for self-service analytics. A commonly quoted axiom is that "big data is for machines; small data is for people".